Beyond Clay Liners: Investigating Shear Strength of Geosynthetic Landfill Systems

Mahima

Contents

1. Introduction..1

    1.1   Background of the problem...1

    1.2   Justification 2

    1.3   Research objectives 4

    1.4   Scope and limitations of the study 4

    1.5   book outline..5

2. Landfill: A general overview 6

    2.1   Introduction 6

    2.2   Landfill legislation in South Africa..6

    2.3   Typical landfill components...9

        2.3.1   Landfill base lining systems 9

        2.3.2   Leachate collection, leakage detection and leachate removal 11

        2.3.3   Landfill capping systems 14

    2.4   Geosynthetics in landfill applications 16

2.4.1   What are geosynthetics? 16

2.4.2   Types of geosynthetics...16

2.4.3   Properties of geosynthetics 17

2.4.4   Application of geosynthetics in landfill system...23

3.   Landfill stability...32

3.1   Introduction...32

3.2   General modes of failure in landfills..32

3.3   Historical landfill failures 35

3.4   Stability of landfill lining system...37

3.5   Methods of stability analysis..38

3.5.1   Limit equilibrium 38

3.6   Factor of safety from slope stability data...45

4.   Review of interface shear strength between geomembrane-geosynthetic combinations....46

4.1   Introduction...46

4.2   Direct shear test experimental approach 46

4.2.1   Different test arrangements...51

4.2.2   Use of gripping and clamping systems..52

4.3   Mobilisation of shear stresses 57

4.4   Interface shear stress-displacement relationship..62

4.4.1   Pre-peak 62

4.4.2   Strain softening..67

4.5   Shear stress versus normal pressure relationship 71

4.5.1   Adhesion 71

4.5.2   Interface friction angle...73

4.6   Effect of material structure on interface friction characteristics.........................79

4.6.1   Effect of geomembrane/ geosynthetic interaction mechanism 80

4.6.2 Effect of geomembrane composition 83

4.7 Summary of previous research 84

4.7.1 Key points 84

4.7.2 Gaps in knowledge 85

5. Research materials and methodology 87

5.1 Introduction 87

5.2 Research materials 87

5.2.1 Geosynthetic Clay Liners 87

5.2.2 Geotextiles 89

5.2.3 Geogrid 91

5.2.4 Geomembranes 92

5.3 Test apparatus and equipment 94

5.4 Research procedures 95

5.4.1 Test sample preparation 97

5.4.2 Material and apparatus setup 98

5.4.3 Testing procedure 103

5.4.4 Testing program 105

5.5 Data analysis 106

5.5.1 Vertical applied stress 106

5.5.2 Resisting shear stress 107

5.5.3 Mohr-Coulomb failure envelope 107

5.6 Quality assurance 108

5.6.1 Repeatability of results 110

6. Results and discussions 113

6.1 Introduction 113

6.2 Shear stress - horizontal displacement relationship 113

6.2.1   Effect of geomembrane composition on shear stress..113

6.2.2   Effect of geomembrane composition on vertical displacement.........................131

6.3   Shear stress – normal stress relationship...134

7.   Practical design application 142

7.1   Introduction 142

7.2   The proposed design methods 142

7.3   Design example 143

7.4   Design solutions 146

7.4.1   Base liner system 146

7.4.2   Capping system...153

8.   Conclusions and recommendations 156

8.1   Summary of conclusions 156

8.2   Recommendations 158

References..160

1. Introduction

1.1 Background of the problem

Solid waste landfills serve as municipal waste containment facilities. Presently, landfills are required to accommodate as much volume of waste as possible while isolating the waste from surrounding air, soil and ground water. With the intention to provide sufficient space for the increasing waste volumes, modified landfill designs have been implemented which incorporate designs with steep side slopes. Although the National Norms and Standards for Disposal of Waste to Landfill provide the minimum requirements and state that alternative design layouts for landfill slopes exceeding 1H:4V may be considered, the new implementation could increase the risk of landfill instability (Department of Environmental Affairs, 2013a; Emery, 2014). Research has shown that instability problems may be associated with shear failure (Russell, et al., 1998; Kim, 2006; Feng, et al., 2010; Duffy, 2016). In such cases, the safety of these containment facilities depends on the shear strength properties of the landfill boundary conditions and construction materials. As a result, proper understanding of shear strength parameters is essential in ensuring safe designs.

A widely accepted construction method for waste landfills in South Africa involved the use of a protective liner system that usually consisted of several layers of geosynthetic materials and compacted clay (Rouncivell, 2007; Dookhi, 2013). These liners can be typically placed on the bottom, side slopes and covers of landfills to improve the stability of the in-situ soil by providing reinforcement, contributing towards leachate drainage, acting as barriers and preventing waste contaminants from leaking into surrounding soil and ground water (Triplett & Fox, 2001; Feng, et al., 2010). The most common installation combinations found in landfills in South Africa and internationally are both soil-geosynthetic and geosynthetic-geosynthetic interfaces; namely geomembrane-soil, geomembrane-geotextile, geomembrane-Geosynthetic Clay Liners (GCL) and GCL-soil (Bergado, et al., 1997; Shukla & Yin, 2006; Qian, 2008a, Dookhi, 2013).

The use of a geosynthetic composite on a slope requires an in-depth knowledge of the friction behaviour of each interface material in contact. The bond developed between a geosynthetic and another geosynthetic or soil should be sufficient to prevent the materials from sliding over each other. A slope lining system placed at an angle, at, or greater than the limiting value at which the soil and geosynthetic materials are able to resist shear could lead to structural failure. A number of geosynthetic interface frictional side slope failures of landfills have taken place around the world (Mitchell, et al., 1990; Reddy & Basha, 2014). An example was a slope stability failure that occurred at Kettleman Hills Hazardous Waste Landfill in California which failed in 1987 due to sliding along interfaces within the composite, multi-layered geosynthetic-compacted clay liner system (Mitchell, et al., 1990). In this landfill failure, the most critical contact surfaces were determined to be those between High Density Polyethylene (HDPE) geomembrane-geotextile, HDPE geomembrane-geonet and HDPE geomembrane-saturated compacted clay (Mitchell, et al., 1990).

The HDPE geomembrane interface instability may have been greatly affected by the interface engineering properties of the geosynthetic and natural materials used. The behaviour of each interface could be different depending on the interactions of the materials in contact. The use of geosynthetic materials that result in high shear strength properties may reduce instabilities of geosynthetic composite slopes.

Understanding the interface friction properties of materials for the design of lined landfills remains crucial, as it is a critical factor governing the integrity and stability of the structure. In order to determine the parameters that govern shear strength, namely angle of friction and cohesion, laboratory or field tests are required.

1.2 Justification

There have been several comparison studies investigating interface friction characteristics of HDPE geomembranes with various other types of geomembranes (Such as Polyvinylchloride (PVC), Chloro-Sulphonated Poly-Ethylene (CSPE), Very Flexible Polyethylene (VFPE) and Ethylene Propylene Diene Monomer (EPDM)) as summarised by Bhatia & Kasturi (1995) and Hillman & Stark (2001). However, only a few studies are available in literature that investigated a comparison between HDPE and Linear Low Density Polyethylene (LLDPE) geomembranes.

Though the design of base liners in South African landfills strictly specify that HDPE geomembranes must be used based on chemical compatibility, long-term durability and ultraviolet (UV) stability of the liner, there is some choice on the type of geomembrane liner that may be used for landfill cap designs (Department of Environmental Affairs, 2013c).

There are several geomembranes being produced and are in use in current practice making it difficult to research on all geomembrane possibilities. Research involving LLDPE geomembranes has been considerably less than that involving HDPE geomembranes because HDPE geomembranes are seen to offer excellent performance for both landfill liner and cover applications while LLDPE geomembranes are only suitable for lining of caps of landfills. LLDPE geomembranes exhibit significantly better extension properties than HDPE geomembranes of the same nominal thickness which can result in high interface friction values (Simpson & Siebken, 1997). Therefore, the differences between HDPE and LLDPE geomembrane interface parameters deserve thorough investigation.

In light of this, a study on the interface frictional parameters produced by geomembranes manufactured from HDPE and LLDPE polymers is carried out in this work. A comparison between two of the most frequently used polyethylene polymers in landfill cover applications was drawn.

Interface parameters from interaction planes, between geomembranes and other geosynthetics found in landfills, are crucial because they are potential surfaces for slope failure. Thus, the results of this study were anticipated to have a valuable contribution towards the selection of particular materials (between HDPE and LLDPE geomembranes) used in industry for landfill cover use. Since there is no one single geomembrane which is suitable for every containment application, this study was conducted to assist in insuring that the most appropriate suggestion and effective material for a containment project was selected. In addition, the results were predicted to increase confidence in the geomembrane products investigated and to reassure engineers of the type of geomembranes they recommended in design projects.

1.3 Research objectives

The main purpose of this research was to determine the effect HDPE and LLDPE geomembrane polymer type has on interface friction characteristics. It was desirable to compare direct shear test results obtained for LLDPE geomembrane with those for HDPE geomembrane in terms of interface shear parameters under different loading conditions. By comparing the friction parameters obtained from both tests, identification of which geomembrane product would provide the anticipated benefits was achieved. In addition, some clarification about the competence of one geomembrane over the other was determined.

In order to achieve the main objective, the specific objectives were:

a) To establish how the structure of the LLDPE and HDPE geomembranes at a micro-level influenced the interface shear behaviour at a macro-level. The study presents an evaluation of the effect of normal stress on the liner system and on interface properties.

b) Conduct detail stability analysis study of various configurations of landfill liner systems to determine whether similar performance patterns exist between HDPE and LLDPE geomembrane friction parameters. Factor of safety values were examined at varying waste depths.

1.4 Scope and limitations of the study

Taking into account the wide range of geosynthetic combinations possible, only a few interfaces among the more common ones were experimented. The investigation conducted was limited to matters that influence interface shear parameters of the HDPE and LLDPE geomembranes tested against various geosynthetic materials i.e. GCLs, geotextiles and geogrid. Direct shear tests conducted were restricted on these geosynthetics and no other type of geosynthetic materials were included in the research. However, the test method proposed, is extendable to new interfaces.

Both types of geomembranes were chosen such that they have the same asperity height and texturing pattern but different compositions which enabled the study to maintain the remaining geomembrane properties constant.

The geosynthetic combinations investigated were based on South African landfill liner system designs as indicated in the South African Waste Information System standards (Department of Environmental Affairs, 2016), by geosynthetic samples sourced both locally and internationally.

Detailed calculations of settlement predictions of landfill foundation materials and solid waste were not included in this book.

1.5 book outline

Chapter 1 introduces the background to the study and provides the reader with the importance of the investigated problem. Chapter 2 provides a general overview of landfill design and the use of geosynthetics in landfill applications. This was followed by Chapter 3 that focuses on landfill stability issues and highlights common methods of stability analysis. A literature review of previous research performed on different geosynthetic-geosynthetic interfaces was presented in Chapter 4.

In Chapter 5, the research materials used in this investigation are described as well as the methodology adopted from the testing standards. The analysis and discussion of the large direct shear test output parameters (i.e. interface friction angle and apparent adhesion) for the HDPE and LLDPE geomembranes sheared against various geosynthetics was detailed in Chapter 6.

Chapter 7 considers the application of the results generated from this study and demonstrates a practical design example of a landfill base and cover lining system evaluated using limit equilibrium analysis methods. Finally, in Chapter 8 conclusions are drawn on the basis of the findings and recommendations are made for future liner implementation and possible further research.

2. Landfill: A general overview

2.1 Introduction

This chapter gives details of the landfill design process by investigating the landfill legislation in South Africa, waste management system and waste disposal in landfill regulations. In addition, the chapter presents landfill containment barrier systems and highlights the leachate collection, leakage detection and leachate removal systems. Furthermore, the chapter explores the different types of geosynthetics available which is followed by an illustration of the physical and mechanical properties of these materials. In conclusion, interest is drawn to the application of geosynthetics specifically looking at the fundamental mechanism of geosynthetic inclusion in landfills.

2.2 Landfill legislation in South Africa

In the past, landfills were large excavations in the ground where household waste was either incinerated or dumped in heaps (Department of Water Affairs and Forestry, 1998b; Munawar & Fellner, 2013). Often, contaminated water, also known as leachate, would be generated from liquid squeezed out of the waste or by rain water that infiltrated into the excavated landfill and saturated the waste (Shukla & Yin, 2006). This toxic liquid seeped into the ground which resulted in ground water pollution. Over the years, this contamination led to environmental safety concerns that influenced the implementation of regulated modern landfills.

Modern landfills are engineered waste containment structures designed to prevent waste and leachate from polluting the surrounding ground water, soil and atmospheric environment. The planning of a landfill must keep harmful pollutants (such as gas and leachate) out of the environment by integrating a drainage layer, gas monitoring system, leak detection system, leachate collection, leachate removal and treatment system (Department of Water Affairs and Forestry, 1998b).

Engineered waste containment facilities use controlled methods of waste disposal. These modern landfills are regulated using waste management guidance documents. Now, all waste disposal sites in

South Africa require that no person may dispose waste unless under the authority of a permit issued in terms of Section 20(b) of the National Environmental Management: Waste Act, 2008 (Act No. 59 of 2008).

The South African government formulated a National Waste Management Strategy in 2011 (aligned with the National Environmental Waste Act (No. 59 of 2008)) which outlined goals to promote waste minimisation through reduction in waste production, re-use of waste for other purposes and recycling of waste before disposing into landfills. In spite of the waste minimisation strategies, large amounts of waste is still being disposed in landfills.

The Waste Classification and Management Regulations define the categories of waste that apply to landfills as depicted in Figure 2-1. Guidance on landfill design and construction detailed in the Norms and Standards for the Assessment of Waste for Landfill Disposal lays down the use of a risk assessment process (shown in Table 2-1) to determine the nature of the waste and the performance requirements of the landfill lining system depending on the risk posed by the disposed waste (Department of Water Affairs and Forestry, 1998c; Zanzinger, et al., 2012). The waste type disposed in a given landfill is an important aspect applicable to the design of the base lining and final capping systems.

Waste Classification	
Hazardous waste (Classified in terms of SAN 10234)	**General waste**
Physical hazards	- Municipal waste (including household hazardous waste)
- Explosives - Flammable gasses, aerosols, liquids, solids or substances - Oxidising gases - Gases under pressure - Self-reactive substances and mixtures - Pyrophoric substances - Self-heating substances and mixtures - Substances and mixtures that, on contact with water, emit flammable gases - Oxidising substances and organic peroxides - Corrosive substances	- Business waste (not containing hazardous waste or hazardous chemicals) - Non-infectious animal carcasses - Garden waste - Waste packaging - Waste tyres - Building and demolition waste not containing hazardous waste or hazardous chemicals - Excavated earth material - Healthcare risk waste (HCRW) - Asbestos waste - Electronic waste (e-waste) - Waste batteries - Putrescible waste
Health hazards	
- Acute toxicity - Skin corrosion and skin irritation - Serious eye damage and eye irritation - Respiratory sensitisation and skin sensitisation - Germ cell mutagenicity - Carcinogenicity - Reproductive toxicity - Specific target organ toxicity – single exposure - Specific target organ toxicity – repeated exposure - Aspiration hazards	
Hazards to the aquatic environment	
- Acute and chronic aquatic toxicity	

Figure 2-1: Waste classification of hazardous and general waste (EScience Associates, 2009;Costley, 2013)

Table 2-1: Department of Environmental Affairs waste classification (Department of Environmental Affairs, 2013a)

Waste type	Disposal requirements	Listed waste types
0 - Very high risk	May not be landfilled. Waste must be treated and retested to determine the risk profile for Class A disposal site classification	N/A
1 - High risk	Class A disposal site. May be disposed of in a H:h/H:H landfill	Asbestos waste, expired, spoilt or unusable hazardous products, PCBs, general waste and domestic waste
2 - Moderate risk	Class B disposal site. May be disposed of in a GLB+ landfill designed in accordance to the requirements specified in the Minimum Requirements for Waste Disposal by Landfill (2nd Ed., DWAF, 1998)	Domestic waste, business waste not containing hazardous waste or hazardous chemicals, non-infected carcasses and garden waste
3 - Low risk	Class C disposal site. May be disposed of in a GLB+ landfill designed in accordance to the requirements specified in the Minimum Requirements for Waste Disposal by Landfill (2nd Ed., DWAF, 1998)	Post-consumer packaging and waste tyres
4 - Inert waste	Class D disposal site. May be disposed of in a GSB- landfill designed in accordance to the requirements specified in the Minimum Requirements for Waste Disposal by Landfill (2nd Ed., DWAF, 1998)	Building and demolition waste and excavated earth
Non-hazardous waste	Class B disposal site. May be disposed of in any	N/A

2.3 Typical landfill components

2.3.1 Landfill base lining systems

A site-specific basal lining system is designed and installed along the base and side slopes of a landfill. Basal lining systems are complex multi-layered systems of soils and geosynthetics constructed on in-situ soil before waste is deposited.

The Norms and Standards for Disposal of Waste to Landfill ensure that disposed waste complies with the allocated landfill type/class. Landfill types are no longer managed in terms of general and hazardous landfills but are now rather based on risk posed as specified in Table 2-1 (Department of Environmental Affairs, 2013a). Depending on the risk, different standard containment barrier requirements are allocated. The varying landfill containment barrier design requirements are separated into classes A, B, C and D as indicated in Figure 2-2.

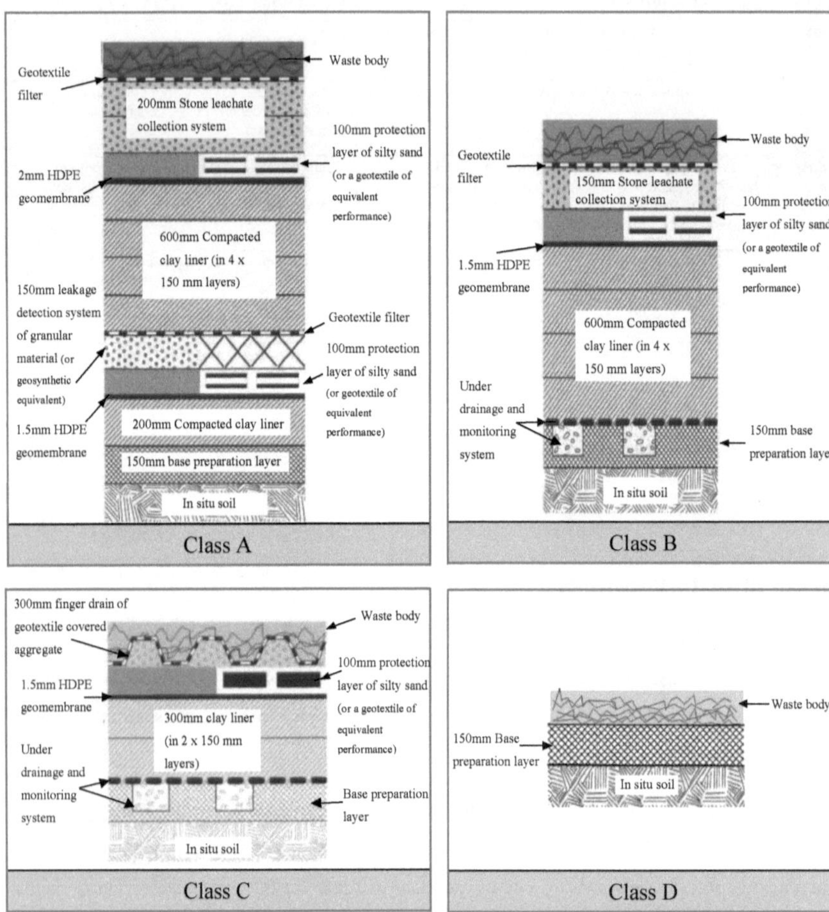

Figure 2-2: Various landfill containment barrier designs (Department of Environmental Affairs, 2013c)

Landfill sites for hazardous waste are classified as either H:H or H:h which reflects the type of waste it may accept. H:h landfills can only accept waste with hazard waste rating of 3 and 4. A H:H can accept all four waste ratings i.e. Waste type 1, 2, 3 and 4 (Department of Water Affairs and Forestry, 1998a). These regulations apply uniformly in all Provinces of the Republic of South Africa (Department of Environmental Affairs, 2013b).

2.3.2 Leachate collection, leakage detection and leachate removal

A leachate management system is required for landfills where significant leachate is generated which is mainly in landfills able to contain waste Type 1, 2 and 3. The planning of a solid-waste landfill must keep harmful leachate out of the environment by integrating a leachate collection, leachate removal, treatment system and a leak detection system in cases of a double liner system (Department of Water Affairs and Forestry, 1998b).

In all landfills, the profile and base of the landfill must be sloped so that any leachate formed is gravitationally directed to a controlled low point (Department of Water Affairs and Forestry, 1998b). Grades of 2% or higher can be designed and constructed for sites with no water table restrictions while for sites with high water tables the design usually has 0.5 to 1% slopes (Koerner, 2005).

The low point of the leachate collection system is equipped with suitable drains or collection pipes that lead to and terminate at a collection point or sump with an outlet leading from this location to beyond the landfill (Department of Water Affairs and Forestry, 1998b; Koerner, 2005). From this low point, the leachate is continuously removed and collected for treatment. Thus, accurate grading of the bottom of the landfill is very important to avoid leachate ponding above the barrier system which can eventually diffuse through it (Koerner, 2005). Figure 2-3 gives some possible low points, contours for gravity flow drainage and collection points in landfills.

There are three approaches toward the removal of collected leachate when it reaches the down gradient sump as illustrated in Figure 2-4: (1) gravity flow, (2) vertical manholes and (3) sidewall risers.

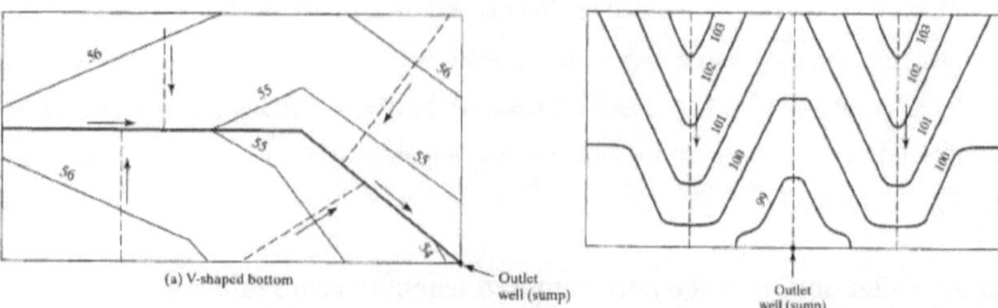

Figure 2-3: Example illustrating possible landfill low points, contours and collection points (Koerner, 2005)

Irrespective of the method of leachate removal, it must be collected and properly treated to a quality standard that is acceptable to the Department of Water Affairs and Forestry before release into the local waterway or sewer system (Koerner, 2005; Department of Water Affairs and Forestry, 1998b). The leachate treatment system could be on-site chemical, physical or biological treatment and/or off-site treatment dependant on the leachate composition (Department of Water Affairs and Forestry, 1998b).

A leachate detection system is required for landfills that are able to contain waste Type 4. Leachate detection systems comprise of elementary liners which are designed to intercept any leachate that passes the barrier of the upper liner in a double liner system. Finger and toe drains are used to direct the leakage to separate leakage collection sumps where the quantity and quality can be monitored, and from which accumulated leakage can be removed. This system allows for 'early warning' monitoring of leachate leaks (Department of Water Affairs and Forestry, 1998a; Department of Water Affairs and Forestry, 1998b).

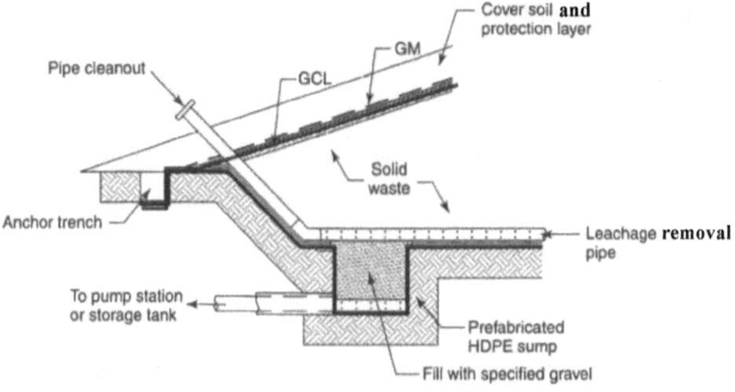

(a) Leachate removal by gravity flow from the bottom of sump

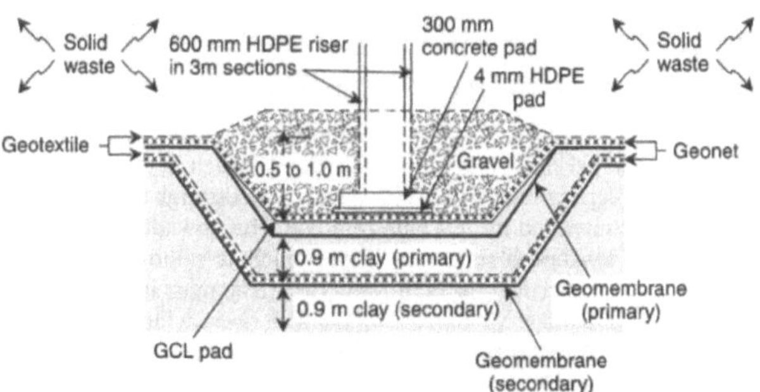

(b) Leachate removal from vertical manhole extending up from sump

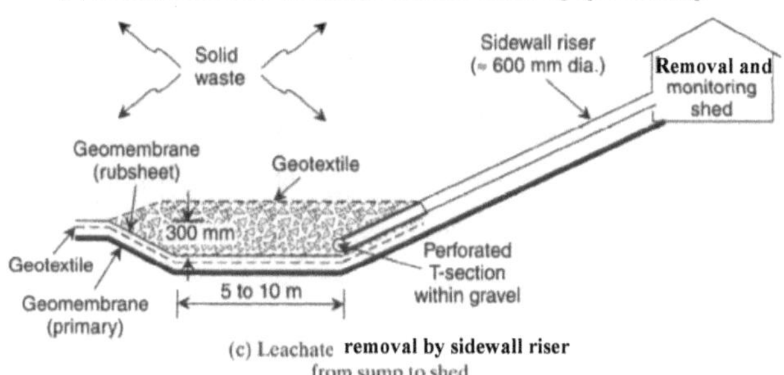

(c) Leachate removal by slope wall riser from sump to shed

Figure 2-4: Various leachate removal designs for primary leachate collection systems (Koerner, 2005)

2.3.3 Landfill capping systems

The Minimum Requirements for Waste Disposal by Landfill (1998b) require that once the landfill has reached its maximum capacity, a landfill capping or cover system is installed over the waste to separate the waste body from the atmospheric environment. This system limits rainwater infiltration into the landfill so that the amount of leachate generated after closure can be reduced (Department of Water Affairs and Forestry, 1998b; Federal Remediation and Technologies Roundtable, 2003).

The design life of a cover system depends primarily on the nature of the waste, the site hydrology and the length of time that the maintenance of the cover will be provided (Legg, 2016). The system is designed to function with minimum maintenance requirement with surface water drainage that prevents water ponding, erosion and instability, as well as robust cover materials that accommodate settling and localised subsidence phenomena (Department of Water Affairs and Forestry, 1998b; Shukla & Yin, 2006; Legg, 2016).

Figure 2-5 demonstrates examples of a landfill cover barrier system with modifications based on the base liner system installed. The components of a cover comprise a combination of some or all of the following layers depending on pollution status of the site and the long term environmental risks posed by the waste (Department of Water Affairs and Forestry, 1998b; Federal Remediation and Technologies Roundtable, 2003; Legg, 2016):

- A vegetation layer which is as thin as possible but sufficient to support shallow rooted plants that have low nutrient needs. This layer is known to minimise erosion problems,
- The soil layer supports the vegetation layer and protects underlying layers. This soil covering is typically 200 mm thick and can consist of crushed stone or cobbles depending on the environment,
- A 900 mm protection layer comprised of local soils or cobbles. Although, this layer is not always included it can help to stop burrowing animals and deep roots,
- A filter layer can be used to prevent fines from the soil layer from clogging the underlying drainage layer. A geosynthetic filter fabric or 300 mm sand layer can provide filtration,

- A drainage layer can be made up of a 300 mm thick stone layer with a low permeability synthetic material. This combination drains by gravity to collector pipes which minimises ponding of water on the geomembrane liner,

- A barrier layer with a low hydraulic conductivity value (e.g. Geomembrane, GCL and/or compacted clay) can create a hydraulic barrier and prevent infiltration of water into the waste. This allows the waste body to be isolated from the surrounding environment (both air and water environments) and

- The gas vent layer prevents uncontrolled escape of landfill odours, methane or toxic gas. Similar to the drainage layer, it consists of 300 mm sand layer or equivalent synthetic material connected to horizontal venting pipes.

Closed landfills can provide important benefits to the surrounding communities and environment. Examples of final practical end-uses of closed facilities can include developing the land to be used as wildlife habitats, parks, sports fields or golf courses.

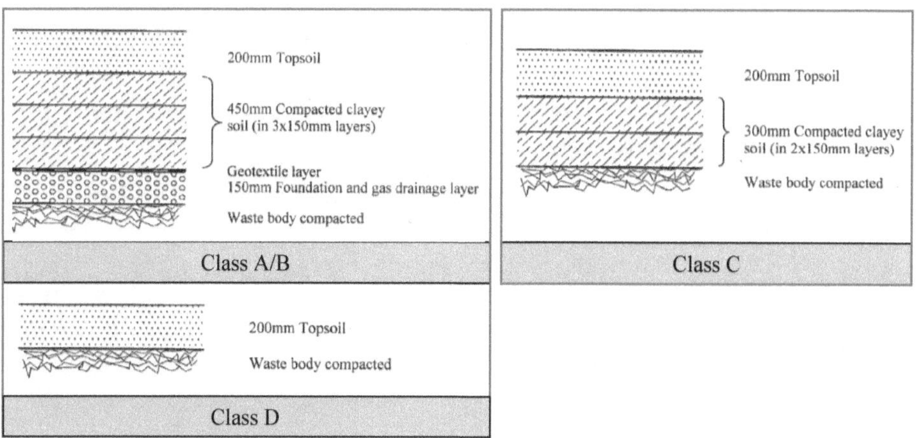

Figure 2-5: Typical cross sections of landfill capping systems (Department of Water Affairs and Forestry, 1998b)

2.4 Geosynthetics in landfill applications

2.4.1 What are geosynthetics?

A geosynthetic is a planar product manufactured from polymeric material used in contact with soil, rock, earth or other geotechnical engineering related material as an integral part of a civil engineering project, structure or system (ASTM D4439). These polymeric materials are known to be able to make impossible designs and applications possible (Koerner, 2005). They can be used in transportation, geotechnical, environmental, hydraulics and private site development designs. Due to geosynthetics' factory quality controlled properties, they are considered more reliable construction materials when compared to their raw material counter-parts. Added advantages are that geosynthetics can be installed rapidly and are generally cost competitive against the construction materials that they replace.

2.4.2 Types of geosynthetics

The planer products manufactured from polymeric materials include; geotextiles, geogrids, geonets, geomembranes, geosynthetic clay liners (GCL), geopipes, geofoams, geocells and geocomposites. These geosynthetic products and their primary functions are illustrated in Table 2-2. Table 2-3 summarises the polymers that are most commonly used to manufacture the various geosynthetics as adapted from Koerner (2005) and Shukla & Yin (2006).

Table 2-2: Primary functions for different types of geosynthetics (Koerner, 2005; Shukla & Yin, 2006)

Type of geosynthetic	Abbreviations	Separation	Reinforcement	Filtration	Drainage	Containment
Geotextile	GT	X	X	X	X	
Geomembrane	GM					X
Geogrid	GG		X			
Geosynthetic Clay Liner	GCL					X
Geonet	GN				X	
Geocomposite	GC	X	X	X	X	X
Geopipe	GP				X	
Geofoam	GF	X				

Table 2-3: Polymers that make up different geosynthetics according to Koerner (2005) and Shukla & Yin (2006)

Polymer material	Abbreviations	GM	GT	GG	GN	GP	GF
High density polyethylene	HDPE	X	X	X	X	X	
Linear low density polyethylene	LLDPE	X	X	X			
Polypropylene	PP	X	X	X		X	X
Polyvinyl chloride	PVC	X				X	X
Polyester	PET		X	X			

2.4.3 Properties of geosynthetics

The properties of geosynthetics vary from physical, mechanical, hydraulic and degradation. The physical properties are associated with the structure, thickness and stiffness of the geosynthetic. Mechanical properties are important in selection, design and use of the material in different civil engineering applications. The mechanical properties that are of concern are loading, shearing and deformation of the geosynthetic. Degradation, abrasion and damage are likely to occur during construction especially at some point in the installation phase.

2.4.3.1 Geogrids

Geogrids are plastics formed into a grid-like configuration. They have large apertures or openings between individual ribs in the machine and cross machine directions. These openings are large enough to allow for soil communication. Geogrids vary considerably insofar as their physical, mechanical and endurance properties.

The primary function of geogrids is reinforcement. They can be used in the reinforcement of a soil for walls, steep slopes, roadway bases and foundation soils. The ribs of geogrids are often quite stiff compared to fibres of geotextiles to be described in Section 2.4.3.3 (Koerner, 2005).

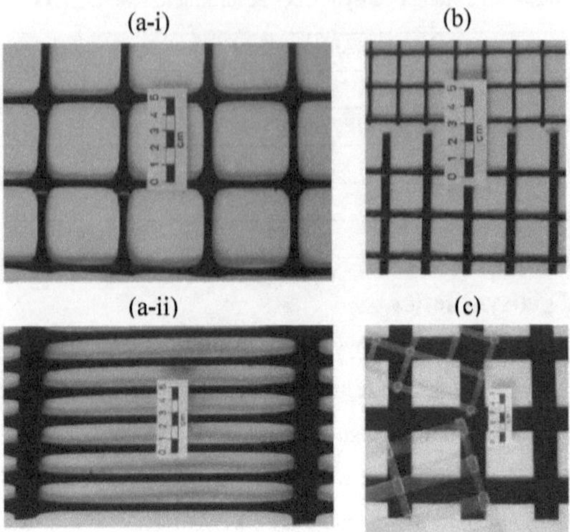

Figure 2-6: Typical geogrids (a) Stiff extruded - (i) Biaxial; (ii) Uniaxial; (b) Flexible woven; (c) Flexible bonded (Shukla & Yin, 2006)

Geogrids can be relatively flexible or stiff depending on the polymers used during production, some examples are shown in Figure 2-6. Stiff geogrids are referred to as unitized or homogeneous geogrids. They have punched holes into High Density Polyethylene (HDPE) and Polypropylene (PP) sheetings, forming a regular pattern. Flexible geogrids are made from high-tenacity polyester yarns, woven into an open structure with the junctions being knitted together or physically intertwined. These flexible geogrids are often coated with PVC, latex or bitumen which assists with dimensional stability and provides protection for the geogrid ribs during soil backfilling (Koerner, 2005).

2.4.3.2   Geosynthetic Clay Liners

Geosynthetic Clay Liners (GCLs) are a combination of polymer and natural soil materials. They are factory-manufactured geocomposite materials that consist of sodium bentonite clay supported by one or two layers of geotextile or geomembrane. The structural integrity of the composite is obtained by what mechanically holds the materials together; needle-punching, stitching or physical bonding illustrated in

Figure 2-7 (Koerner, 2005). Reinforced GCLs (stitch-bonded) are best suited for applications involving low normal stresses (e.g. pond and lagoon liners and cover systems); whereas needle-punched GCLs are the better choice for applications where a high normal stress is applied (e.g. landfill bottom liners). The use of unreinforced GCLs is not recommended for slopes steeper than 10:1 (Shukla & Yin, 2006).

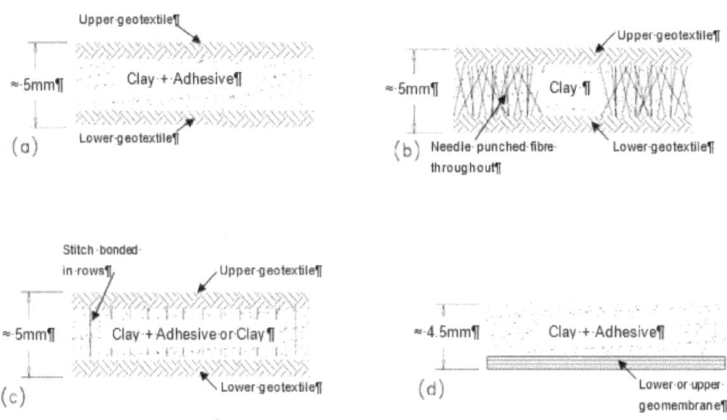

Figure 2-7: Types of geosynthetic clay liners (a) Unreinforced with upper and lower geotextile; (b) Needle punched; (c) Stitch bonded; (d) With geomembrane (Shukla & Yin, 2006)

GCLs are used as hydraulic barriers against water, leachate or other liquids and sometimes even gases (Koerner, 2005). These polymers are used beneath a geomembrane or by themselves and act as composite liners within landfill bottom liners and cover systems (Koerner, 2005). When GCLs are used in landfills, they provide limited thickness, good compliance with differential settlement of underlying soil or waste, easy installation and low cost.

Stability considerations are very important when GCLs are placed on side slopes because of the very low shear strength of sodium bentonite after hydration which can provide a potential surface for slope failure.

Both internal and interface shear strengths of GCLs and adjacent materials are needed for stability analysis in design (Fox & Stark, 2004).

2.4.3.3   Geotextiles

Geotextiles are fabrics made of flexible synthetic fibres manufactured by standard weaving machinery or are matted together in a random nonwoven manner. They are porous to liquid flow across their manufactured plane and within their thickness and are usually seen as alternatives to granular soil filters. The fabric's functions include: separation, reinforcement, filtration and/or drainage (Koerner, 2005).

There are five principal types of fibres used to manufacture geotextiles: (1) monofilament, (2) multifilament, (3) staple fibre yarn, (4) slit film monofilament and (5) slit-film multifilament shown in Figure 2-8 (Koerner, 2005). These fibres and yarns can be manufactured into fabrics. The basic fabric manufacturing choices are woven, nonwoven or knit as illustrated in Figure 2-9 below. Variations have been found to have direct influence on the physical, mechanical and hydraulic properties of the fabrics (Koerner, 2005).

A web is formed by bonding the fibres and yarns together using one of the following methods: (1) needle punching, (2) resin bonding and (3) melt bonding (Koerner, 2005). Needle punching is the most common nonwoven geotextile bonding method. This method requires hundreds of specially designed needles with three or four downward-oriented barbs, indicated in Figure 2-10, used to reorient the fibres of a fibrous web. The process allows for mechanical bonding to be achieved throughout the length and width of the fabric. In the resin bonding process, a fibrous web is either sprayed or impregnated with an acrylic resin to form bonds between filaments. In the melting bonding process, the continuous filaments or staple fibres web is heated and melted together which results in stiff fabrics (Koerner, 2005).

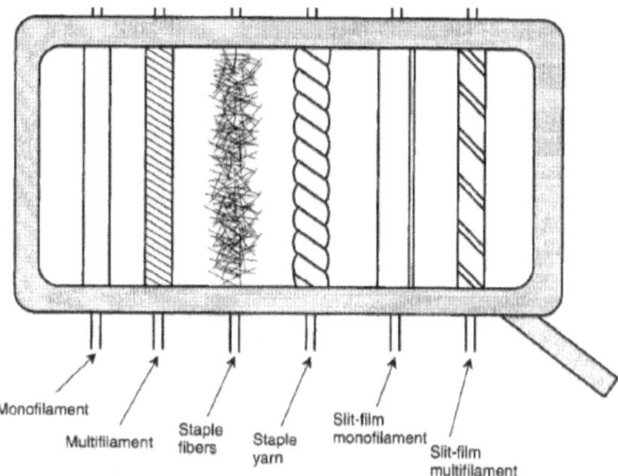

Figure 2-8: Types of polymeric fibres (or yarns) used in the manufacture of geotextiles (Koerner, 2005)

Figure 2-9:(a) Woven monofilament calendered, (b) Woven multifilament, (c) Woven slit (split) film, (d) Nonwoven needle punched and (e) Nonwoven heat bonded (all images have a magnification of X30) (Koerner, 2005)

Figure 2-10: Diagram of details on typical needle punching needles (Innovation in textiles, 2012)

If normal stresses are applied on a geotextile, its thickness will decrease, and compression will occur. Although compressibility is relatively low for most geotextiles, it is an important mechanical property. For geotextiles that convey liquid within the plane of their structure such as nonwoven needle-punched geotextiles, compressibility is imperative. It is important that the geotextile allow for this flow to occur and is not impeded (Koerner, 2005).

2.4.3.4   Geomembranes

Geomembranes are thin impervious sheets of flexible thermoplastic polymeric material with a primary function to contain liquid or solid-storage facilities (Koerner, 2005). This includes all types of landfills liners and covers, reservoirs, lagoon liners, wastewater treatment facilities, canal linings, floating covers, tank linings and other containment systems. These sheets are manufactured in a factory and often seamed on the field (Koerner, 2005).

There are various geomembranes available in the industry but the basic difference between them is the material and/or method of its manufacture. Polyethylene is the most commonly used polymer in the manufacturing of geomembranes (Bhatia & Kasturi, 1995; Koerner, 2005).

Geomembranes are available with smooth and textured surfaces shown in Figure 2-11. The interface friction obtained between adjacent geosynthetics with textured geomembrane surfaces is greater than that obtained with smooth surfaces. This phenomenon greatly improves stability particularly for geomembranes placed on sloping ground (Coronel, 2006; Shukla & Yin, 2006).

Geomembranes are available in different texturing patterns and asperity heights. The asperity creates a roughened; high-friction surface from a smooth geomembrane sheet through a texturing process. The four methods used to texturize geomembranes are; (1) coextrusion, (2) impingement, (3) lamination and (4) structuring (Koerner, 2005). Coextrusion uses nitrogen blowing agents to extrude molten extrudate and create a finished textured sheet. The second method of texturing is impingement; a process in which hot polyethylene particles are actually projected onto the previously manufactured smooth sheet. In the lamination method, a foaming agent contained within molten polyethylene provides a froth that is adhered to the previously manufactured smooth sheet providing a rough textured surface. This method is expensive and is rarely used.

The fourth method and more common is structuring. Smooth hot sheets (approximately at $120\,^\circ C$) are pressed and allowed to deform under two counter-rotating patterned rollers which allows texturing to occur (Koerner, 2005).

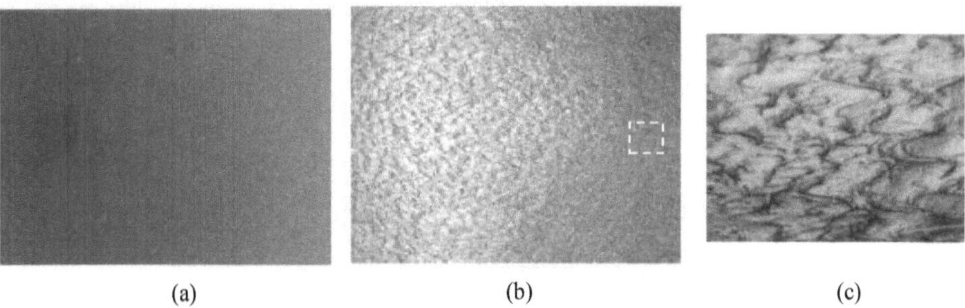

(a)   (b)   (c)

Figure 2-11: Top view of geomembrane samples (a) Smooth; (b) Textured and (c) Detail of textured geomembrane sample (Kim, 2006; Blond & Elie, 2006)

2.4.4 Application of geosynthetics in landfill system

This subsection shows a landfill base and cover lining system example that incorporates majority of the geosynthetics mentioned in Section 2.4.3. The description is followed by an illustration of the benefits of combined geosynthetic and geomembrane inclusion in landfill designs.

2.4.4.1   Benefits of geosynthetics in landfill base liners and covers

The lining of the bottom and sides of landfills is a necessity due to the water in the incoming waste materials which is increased by rainfall and snowmelt interacting with the already placed waste and forming a liquid called leachate as illustrated in Figure 2-12. This leachate flows gravitationally downwards and if not for a liner would continue to flow until it encountered groundwater or surface water posing the threat of pollution (Koerner, 2005). Geomembranes are the most frequently used geosynthetic type in landfill lining systems because they protect surrounding soil and groundwater from contaminated fluids by creating a relatively impermeable liquid barrier when compared to other geosynthetics or soils.

Geomembranes work more effectively as composite liners (i.e. geomembrane on low-permeability soil such as clay). For example, leachate would move easily through a hole in a geomembrane and seepage would take place through the soil subgrade. With a low-permeability soil liner alone, seepage would take place over the entire area of the soil liner. On the other hand, inclusion of a composite liner reduces the rate of leakage because the amount of leachate passing through any hole in the geomembrane encounters low-permeability clay soil, which reduces further migration of leachate passing through the hole (Shukla & Yin, 2006).

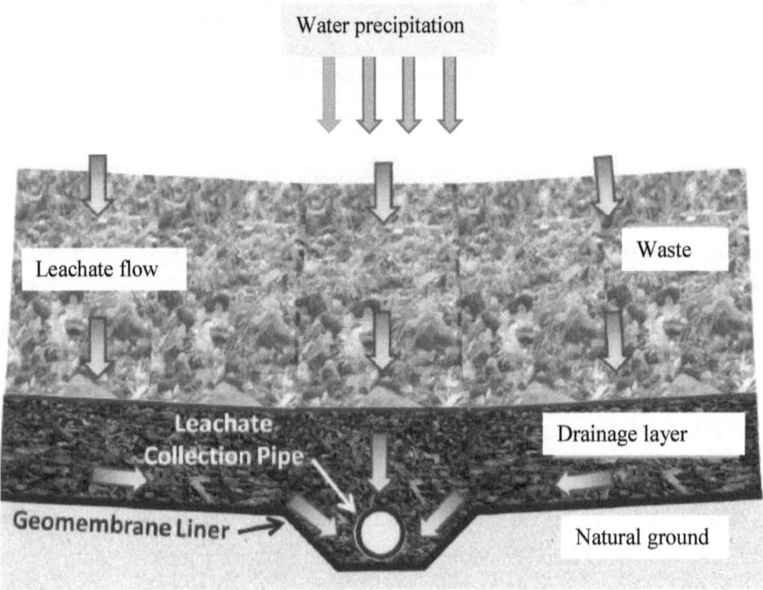

Figure 2-12: Example of leachate generation in landfills due to water precipitation (Mwai, et al., 2010)

In order to minimise or entirely eliminate leachate generation after solid-waste filling is complete, a geomembrane cover is required over a landfill, waste pile or other mass of solid material. Covers are used for various reasons including; reduce air pollution (for landfills holding chemicals and agricultural waste), temperature control for anaerobic decomposition of agricultural wastes and reduce odour (from manure and other biodegradable farm waste) (Koerner, 2005).

The most widely used geomembrane in the waste management industry is High Density Poly-Ethylene (HDPE) (Shukla & Yin, 2006). Flexible geomembranes such as Poly Vinyl Chloride (PVC) or Linear Low Density Poly-Ethylene (LLDPE) are best used in non-uniform surfaces often seen in landfill covers (Shukla & Yin, 2006; Agru, 2016; Contain Enviro Services, 2016).

Polyethylene belongs to a family of materials classified as polyolefins (United States Plastic Corparation, 2008; Contain Enviro Services, 2016). These polyolefins are high molecular weight hydrocarbons and include LLDPE, low density polyethylene (LDPE), HDPE, polypropylene copolymer (PPC), polypropylene (PP) and polymethyl pentene (PMP). The most commonly used of these polymers in the manufacturing of geosynthetics are HDPE, LLDPE, and PP polyolefins (United States Plastic Corparation, 2008). While these polymers have very similar characteristics, their individual chemical make-up gives them different mechanical properties which need to be considered during the geomembrane selection process (Contain Enviro Services, 2016).

The difference between LDPE, LLDPE and HDPE geomembranes is the way their cellular structure or molecules bond is formed (Global plastic sheeting, 2017a). According to the United States Plastic Corparation (2008), polymerized ethylene can result in relatively straight polymer chains with branches. By varying the degree of branching, different kinds of polyethylenes can be achieved. As noted in Figure 2-13, HDPE geomembranes have minimal branching of its polymer chains. The reason for less branching is because HDPE liners are denser, more rigid and less permeable then LLDPE or LDPE. LLDPE geomembranes have a significant number of short branches which enables its chains to slide against each other upon elongation without becoming entangled like LDPE. LDPE polymers have long branching chains that would get caught on each other upon stretching.

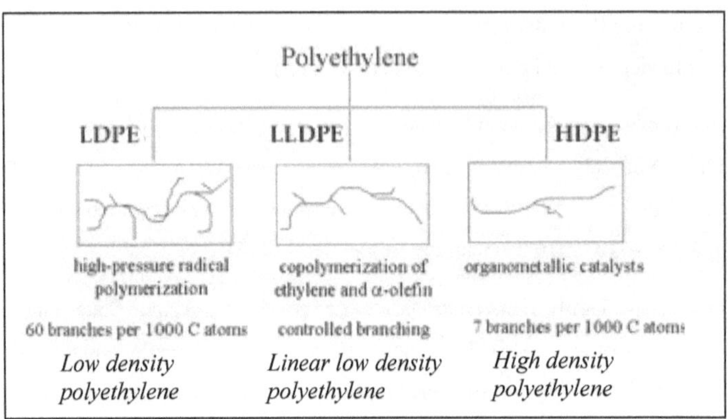

Figure 2-13: Different degree of branching of LDPE, LLDPE and HDPE chemical structures (Arrighi & Kraft, 2011)

HDPE geomembranes have the distinction of being the strongest, toughest, most puncture and chemical resistant and least flexible of the two types of geomembranes (LDPE geomembranes are not common). HDPE is the most UV resistant and does not need UV additive packages (United States Plastic Corparation, 2008). HDPE's strength comes from its tight cell structure that makes it very difficult for other molecules to pass through its structure on a microscopic level (Agru, 2016; Global plastic sheeting, 2017a).

LLDPE have lower molecular weight resin that enables it to be more flexible, have a higher tensile strength and more conformability making it suitable for landfill caps, pond liners, lagoon liners, potable water containment, tunnels and tank linings (Shukla & Yin, 2006). LLDPE geomembranes are also more pliable, softer and can easily conform to non-uniform surfaces. LLDPE are used for geomembranes that need high impact and puncture resistance (often from foot traffic). The molecules all line up and strongly hold together as the geomembrane elongates allowing good environmental stress crack resistance (Global plastic sheeting, 2017a; Global plastic sheeting, 2017b). Although HDPE are typically less expensive than LLDPE, LLDPE liners can be manufactured in thinner grades which may offset this advantage (Contain Enviro Services, 2016).

In landfill applications, LLDPE or HDPE geomembranes are commonly used in conjunction with geosynthetic clay liners (GCLs), geogrids or geotextiles to provide the strongest protection, avoid leakage and groundwater contamination.

2.4.4.2   Benefits of geosynthetic–geosynthetic interactions

Since geomembranes are the most frequently used geosynthetic type in landfill lining systems, they are often used together with other geosynthetics to ensure the landfill lining system functions effectively. Each geosynthetic used should meet certain functional requirements independent of the performance of the other geosynthetics. This subsection illustrates how geotextiles, geosynthetic clay liners and geogrids function with geomembranes in a landfill application.

2.4.4.2.1   Geogrid

Geogrids are often used between geocells (filled with natural soil) and geomembranes in cover systems. The mechanism of interlocking between aggregate particles and geogrid apertures causes an increase in local stiffness through friction and interaction of particles with the geogrid. This occurs through a transfer of stresses from the soil to the geogrid reinforcement made through passive resistance at the soil to the grid cross-bar interface. Owing to the small surface area and large apertures of geogrids, the interaction is mainly expected to be interlocking rather than friction. However, an exception occurs when the geogrid is placed adjacent to a geomembrane. Similar to when the geogrid is placed next to a soil with small particles, the interlocking effect is negligible because no passive strength is developed against the geogrid instead the local stiffness is increased through friction. In a geomembrane/ geogrid interface, the geogrids can reduce the tensile stresses experienced by an adjacent geomembrane or accommodate landfill stability by reinforcing overlying cover soil (Shukla & Yin, 2006).

2.4.4.2.2   Geosynthetic Clay Liner

Geosynthetic Clay Liners (GCLs) are used as a substitute for compacted clay liners in cover systems and composite bottom liners. GCLs function in a similar way to a geomembrane which forms a relatively low hydraulic conductive barrier. The GCL barrier prevents seepage of water, leachate or other liquids and sometimes gases. GCLs are used in combination with geomembranes, geotextiles and geonets (as illustrated by Figure 2-14) to provide redundant safety measures for the landfill design.

GCLs can experience two failure modes when placed on a slope in a landfill lining system; interface or internal failure. One of the main challenges faced when using GCLs is that hydrated bentonite can result in low shear strength characteristics. The shear strength of a GCL for interface and internal failure modes is significantly affected by wet conditions (Bacas, et al., 2015). Reinforced GCLs have a composite structure that utilizes the strength of a strong mechanical bond which can be from millions of needle-punched fibres to secure a uniform layer of sodium bentonite (Fibromat, 2014). When the bentonite layer hydrates, the pores swell and the needle-punched fibres become tensioned thereby forming a watertight sheet that offers protection to the overlying layer i.e. geomembrane liner. The bentonite swells under a confined stress to ensure that a self-sealing barrier forms within a confined defined space. The confining stress increases as the waste in the landfill increases, and the hydraulic conductivity of a GCL generally decreases significantly due to the lower void ratio of bentonite resulting from higher confining stresses (Shukla & Yin, 2006). Therefore, a reinforced GCL only functions properly if hydrated and under a confining stress. When GCL hydration is expected in the field, shear strength tests should be conducted under hydrated conditions to get the "worst" case interface behaviour. Full hydration is always expected in the field unless the bentonite is encapsulated between two geomembranes (Fox & Stark, 2004).

Several investigators have evaluated the GCL internal shear strength using direct shear tests (Zornberg, et al., 2005; Koerner, 2005; Mackenzie & Du Toit 2011). Their investigations have shown that GCL interface shear failure (sheared against textured geomembrane) would occur before GCL internal shear failure. The conclusion is well-supported by published test results from low normal stress up to 1500 kPa. Similarly, a number of authors' databases indicated that failure always occurred at the GCL interface and no internal GCL failures were observed (Triplett and Fox, 2001; Chiu and Fox, 2004; Fox & Stark, 2004; Zornberg, et al., 2005; McCartney, et al., 2009; Lin, et al., 2014).

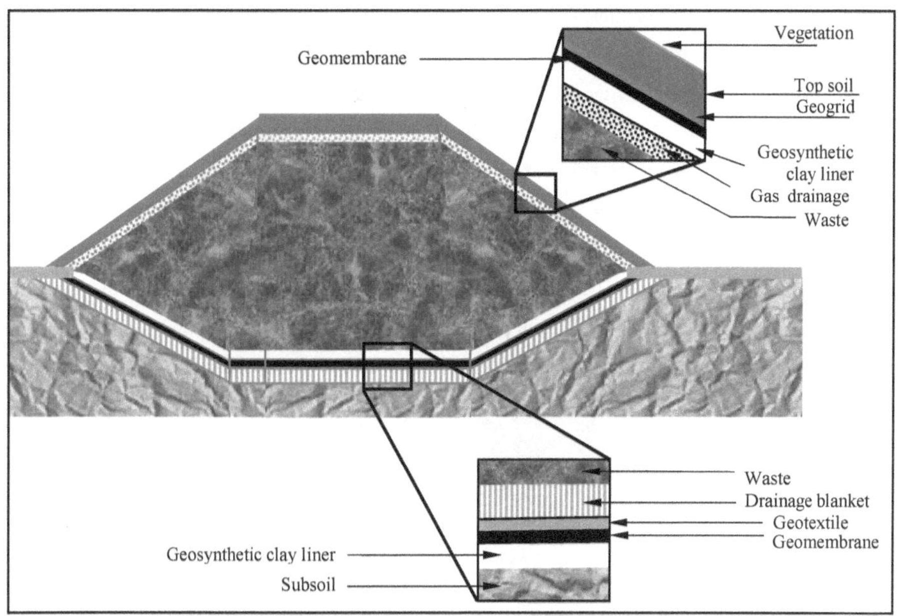

Figure 2-14: Geosynthetic placement in landfill lining system

Some authors found that the friction failure of HDPE textured geomembrane sheared against hydrated needle-punched GCLs (nonwoven side) specimens can change from interface shear to GCL internal shear failure as normal stress increases (Triplett & Fox, 2001; Fox & Ross, 2011). In addition, research conducted at the University of Cape Town established that geomembrane/ textured GCL tests conducted above 36 kPa normal stress resulted in partial or full internal failure of the GCL instead of failure at the desired interface (Rouncivell, 2007). Other researchers concluded that shearing a GCL against a smooth geomembrane will result in internal failure. Currently reported values of GCL internal and interface shear strengths have significant variability due to variation in component materials, differences in manufacturing processes and changes in the design of GCLs (Fox & Stark, 2004).

2.4.4.2.3   Geotextile

One of the functions of geotextiles in landfills is to act as a filter layer between the primary waste and the leachate collection system. When a geotextile is placed adjacent to landfill leachate, a discontinuity arises at the interface between the leachate and the structure of the geotextile. Landfill leachate can contain high suspension solids coupled with high-micro-organism content. It is essential that a condition of equilibrium can be established at the contact interface to prevent micro-organism content found in landfill leachate from combining and building up on or within the geotextile. This process can be prevented by ensuring that geotextiles have adequate permeability which allows fluid movement while limiting the uncontrolled excessive migration of micro-organism content across its manufactured plane over a projected service lifetime. Sufficient permeability is required both across and perpendicular to the geotextile's manufactured plane. Therefore, geotextiles allow adequate liquid flow while ensuring that clogging of the fabric does not occur.

At the leachate/geotextile interface, there is an initial loss of some suspended solids particularly solids adjacent to the geotextile filter and having diameters smaller than the filter pore spaces. These solids migrate through the geotextile openings under the influence of liquid free flow as shown in Figure 2-15. This process occurs until an equilibrium is reached. There are generally three zones that identify the equilibrium: (1) the undisturbed suspended solids, (2) a 'filter' layer which consists of progressively smaller solids as the distance from the geotextile increases and (3) a bridging layer where the remaining larger solids bridge over the pores in the geotextile and retain smaller solids (MIRAFI, 1996; Shukla & Yin, 2006).

The geotextile can also function as a protective layer when it is placed between two dissimilar materials i.e. a geomembrane and angular soil particles a combination commonly found in a landfill (Figure 2-14). In this case, sand and other conventional soil layers are no longer used as a protection layer above a geomembrane (Russell, et al., 1998). As an alternative, nonwoven geotextiles with a high mass per unit area are used as a cushioning layer that prevents puncture for the geomembrane against the gravel drainage layer placed above it (Bhatia & Kasturi, 1995; Shukla & Yin, 2006). The geotextile can alleviate stresses and strains from being transmitted to the geomembrane surface. This allows the integrity and functioning of both materials to remain intact (Koerner, 2005; Shukla & Yin, 2006).

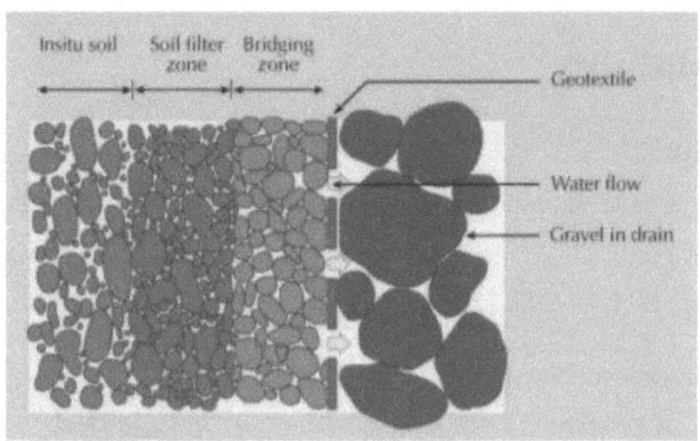

Figure 2-15: Graded soil filter formation adjacent to a geotextile (MIRAFI, 1996)

2.4.4.2.4 Summary

Geosynthetics are polymeric materials used for different geotechnical engineering applications. These planar materials are available in different forms and have different properties which enables them to be used in various civil engineering designs. Applications such as landfill liner systems can incorporate a number of geosynthetics at a time. The most commonly used geosynthetic in landfill systems is a geomembrane. The geomembrane can be used together with several other geosynthetics to ensure the landfill lining system functions effectively i.e. geomembrane/ geogrid, geomembrane/ GCL and geomembrane/ geotextile interfaces. In this document, these interfaces are collectively referred to as geomembrane/ geosynthetic interactions. The geomembrane/ geosynthetic interaction formed between two materials influences the performance of the lining system. The determination of interaction friction characteristics is crucial during the design process which requires the shear strength of each geosynthetic/ geosynthetic interface in a landfill to be known.

3. Landfill stability

3.1 Introduction

The aim of this chapter is to illustrate that engineered landfills are vulnerable to failure. According to several researchers the factors affecting the stability of a landfill include the following (Xuede, et al., 2001; Omari, 2012; Jahanfar, 2014; Dookhi, 2013):

- Geometry (inclination of the slope)
- Shear strength of material
- Loading conditions
- Pore water pressure (leachate level and movement within the landfill)
- Settlement
- Operations

This chapter begins with a description of previous landfill failure case studies and the general modes of failure landfills can experience. The description is followed by an illustration of the different methods available for stability analysis. In conclusion, interest is drawn to acceptable factor of safety values expected in a stability analysis.

3.2 General modes of failure in landfills

Waste containment failure can occur at various stages. For example, failures could occur during excavation, during liner system construction, during waste filling and after landfill closure (Qian, et al., 2003). In addition, landfills can fail in several ways. The two main types of failure methods are the rotational and translational failure. Both failure types can involve the waste mass and even the subsoil because the construction of landfills usually involves excavating into natural soils. An understanding of the various modes of slope failures in landfills are summarised in Figure 3-1.

a) Rotational failure could occur behind the waste mass or beneath the site due to unstable soil mass. Failure can develop along the slope, at the toe, or within the base foundation subsoil. This geotechnical problem is site specific and does not involve the failure of the liner systems or waste properties but only the failure of the subsoil (Pollux waste to energy, 2016; Reddy & Basha, 2014),

b) Soft foundation subsoils can initiate failure and uplift in the soil that can spread up through the lining system and waste mass (Reddy & Basha, 2014),

c) Rotational failure can occur completely within the waste mass without propagating to the liner system. It is handled exactly as the rotational failures through waste, liner and subsoil failure highlighted above. The only difference is that the material is municipal or hazardous waste instead of soil,

d) Sliding failure can occur at great depths where the movement of a layer of waste above the liner system at the base of a landfill can propagate failure either through the waste mass or along the landfill slopes, until it exits at the toe of an adjacent slope (Reddy & Basha, 2014; Duffy, 2016),

e) A combination of heavy rainfall and steep slopes can result in the leachate collection system sliding on the underlying liner system. If this mode of failure occurs sand or gravel can be used onto the lined slope and

f) Similar to how sliding failure of the leachate collection system occurs, the final cover system can slide on the liner system due to heavy rainfall and if the slope is too long or steep.

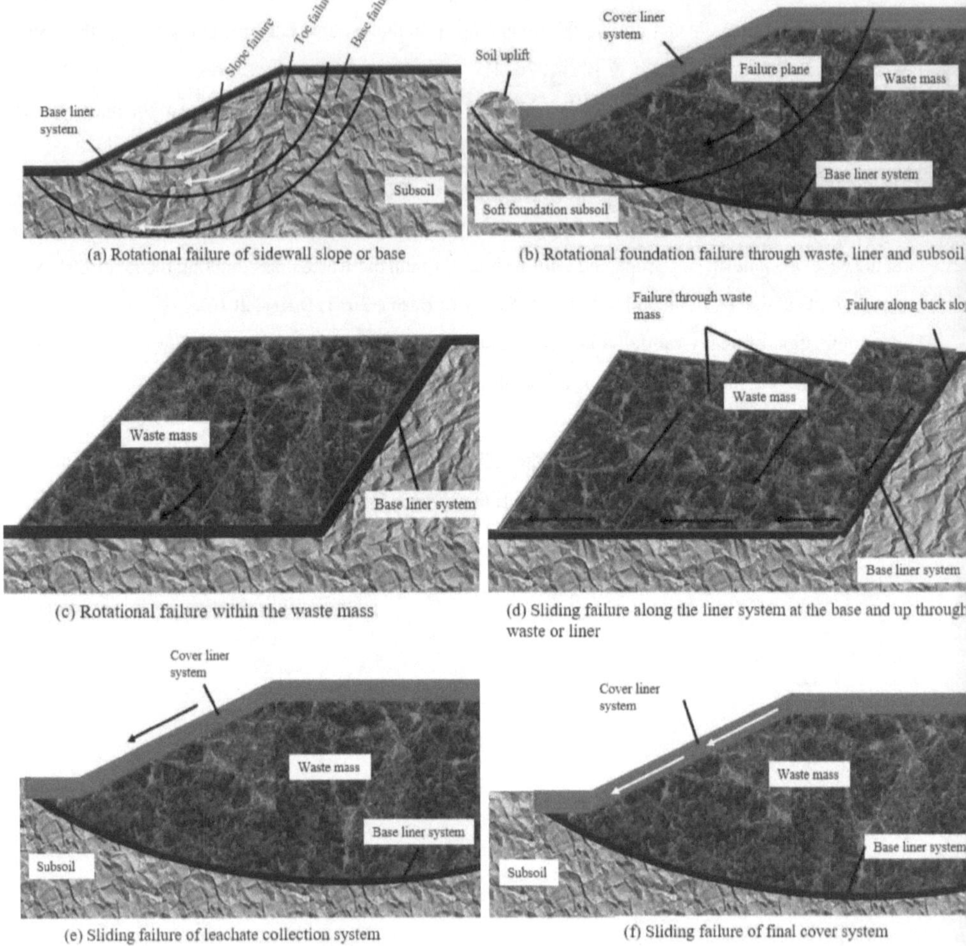

(a) Rotational failure of sidewall slope or base

(b) Rotational foundation failure through waste, liner and subsoil

(c) Rotational failure within the waste mass

(d) Sliding failure along the liner system at the base and up through waste or liner

(e) Sliding failure of leachate collection system

(f) Sliding failure of final cover system

Figure 3-1: Various landfill failure modes

3.3 Historical landfill failures

Several landfill failures have occurred all over the world for various reasons. Most of the events have been described in the literature as sliding waste (e.g. at Umbaniye dump site in Istanbul in Turkey, Rumpke Sanitary landfill in Cincinnati in the United States of America and Leuwigajah landfill in Indonesia (Illustrated in Appendix I)). The Leuwigajah landfill waste slide failure was triggered by water pressure in the soft subsoil (Bauer, et al., 2008; Landva & Dickinson, 2012; Lavigne, et al., 2014). Similarly, the Rumpke Sanitary landfill failure was widely considered to have been a translational slide along a weak layer of native soil directly beneath the bottom of the waste (Landva & Dickinson, 2012). Other authors believe that failure occurred in the waste itself and not in the native soil beneath. It is possible for waste material to decompose to the state of a slime, in which case its strength could become extremely low. This can result in a movement of waste mass over a zone of highly decomposed, mostly organic waste material, which was found to be the cause of failure at the Umbaniye dump site in Istanbul in 1993 (Landva & Dickinson, 2012).

There are several factors that could contribute to landfill failures. The most common factor being a build-up of pore pressures due to increased water in the landfill. Excessive leachate level build-up within old decomposed waste caused by water infiltrating from adjacent surface water ponds creates pore pressures which results in weakened surfaces that trigger failure (Landva & Dickinson, 2012). This was the cause for the Umbaniye dump site failure. When pore pressures are introduced, effective shear strength is reduced to a point less than the true peak shear strength at a surface. Leachate can be reduced and removed using the leachate collection and removal system (Russell, et al., 1998; Shukla & Yin, 2006; U.S Department of Transportation Federal Highway, 2008). Other landfill failures caused by pore pressures reported in literature include; the Rumpke Sanitary landfill in Cincinnati in 1996, Hiriya landfill in Israel in 1997 (illustrated in Appendix I), Bogota landfill in Colombia in 1997, Durban landfill in South Africa in 1997 (shown in Appendix I), Payatas landfill in the Philippines in 2000, Athens landfill in Greece in 2000, Leuwigajah landfill in Indonesia in 2005 and most recently the Xerolakka municipal solid waste landfill failure in Greece in 2010.

The Payatas landfill in the Philippines failed due to pore water pressure build up after heavy rains and low density of waste as shown in Figure 3-2. This failure occurred after two weeks of heavy rain from two typhoons that hit the area. Slope stability analyses indicated that the raised leachate level and a significantly over-steepened slope caused increased pore pressures within the landfill and resulted in a decrease in effective stress and stability (Koelsch, 2000; Kavazanjian Jr. & Merry, 2005; Bauer, et al., 2008; Jafari, et al., 2012). A similar waste slide occurred in Hiriya landfill in Israel and Xerolakka municipal solid waste landfill in Greece. Both of these case histories demonstrated that a combination of heavy rainfall and increased steepening of the landfill slopes could experience slope failure from elevated leachate or high pore pressures (Athanasopoulos, et al., 2013; Reddy & Basha, 2014). Literature has indicated that unanticipated pore pressures conditions may occur in landfills due to clogging of the leachate collection system (Thiel, 2001).

Figure 3-2: Aerial view of Payatas landfill with slope failure and homes visible in foreground (Jafari, et al., 2012)

Researchers have also found that generation of landfill gas pressure from biodegradation of waste can contribute in causing failure. In 2003, an Athens landfill failed which was speculated to have been caused by a fire initiated by landfill gases, which occurred two weeks before the failure (Landva & Dickinson,

2012). Similarly, the Indonesian Leuwigajah landfill failure that ocured in 2005 was most likely triggered by severe damage of reinforcement particles due to a smouldering landfill fire (Bauer, et al., 2008; Landva & Dickinson, 2012; Lavigne, et al., 2014).

3.4 Stability of landfill lining system

According to Qian & Koerner (2014) finding the location of the liner failure interface is the most important issue for the stability analysis of multilayer lined landfills. This has led to various researches to be carried on the shear strength and possible translational failure where the liner components may slide over each other or along the slopes of the landfill. The contact behaviour and interaction between various materials is an important consideration to ensure that weak interfaces do not frequently form within such environmental containment facilities.

The importance of interface strengths was demonstrated by the several translational failures that happened in the past highlighted in Table 3-1. This table illustrates that previous landfill failures have shown two main ways in which interface translational failure can occur: either due to excessive wetness between liner interfaces especially on a compacted clay liner layer or sliding developing along interfaces with low frictional resistance.

Table 3-1: Previous landfill disasters caused by translational failure (Mitchell, et al., 1990; Reddy & Basha, 2014)

Year	Location	Cause of failure
1988	Kettleman, California, USA	Sliding developed along interfaces with low frictional resistance within the multilayered geosynthetic-compacted clay liner system beneath the waste mass. The most critical contact surfaces were determined to be those between HDPE geomembrane and geotextile, HDPE geomembrane and geonet and HDPE geomembrane and saturated compacted clay.
1988	L-1 landfill in North America	A landfill lined with geomembranes, geonets and geotextiles experienced translational failure due to excessive wetness of the compacted clay liner (CCL) and geomembrane interface at the base of the landfill.
1994	L-2 lined landfill in Europe	A section of geomembrane was pulled out of its anchor trench due to translational failure of a wet CCL and HDPE geomembrane interface.
1997	L-3 lined landfill in North America	The landfill composite liner system consisted of a geomembrane, GCL and CCL. Failure was caused by wet bentonite within an unreinforced GCL.

To mitigate landfill interface translational failure, deeper understanding of the characteristics that cause this failure to occur is required. Assessment of failure using a rotational failure method rather than translational failure analysis to estimate the weakest layers is likely to result in an overestimation of the stability, corresponding to a higher factor of safety than may actually exist (Qian, et al., 2003). As a result, Qian, et al. (2003) recommended conventional wedge analysis methods could be used to evaluate potential failure systems for base liner and final capping systems.

3.5 Methods of stability analysis

3.5.1 Limit equilibrium

The internal stability of structures with geosynthetics could be analysed using a number of limit equilibrium design methods. Limit equilibrium methods could be applied to many geotechnical problems and have been the most widely used analytical technique within the context of slope stability analysis (Duncan, 1996). In principle, the methods are all concerned with satisfying boundary conditions, force and/or moment equilibrium and the failure criterion along an assumed slip surface. This failure surface may be a circular or non-circular arc, a logarithmic spiral or any other arbitrary surface.

In South Africa, the limit equilibrium analysis method for landfill liner stability was typically a preferred approach adopted in design practice (Dookhi, 2013). The method provides a reasonable representation of the potential failure surfaces. Therefore, the finite length slope analysis was carried out to identify the most critical interface in terms of stability as shown in Figure 3-4.

3.5.1.1 Base liner system (Basal liner)

As seen in Figure 3-1, translational failure at the base can occur as sliding above, within or beneath the liner system and propagate through the waste or along the lined back slope. A two-part wedge method for translational failure analysis developed by Qian (2008a) can be used to calculate the factor of safety (FS) for a landfill base design of waste placement. Figure 3-3 illustrates a waste mass base configuration demonstrating a reasonable replication of the base lining system. The figure enables the factor of safety for the waste mass against possible translational failure on predetermined sliding failure surfaces to be calculated (Qian, 2008a). The waste mass has an active wedge lying on the back slope that is inclined to cause failure and a passive wedge lying on the landfill's foundation soil or base liner system which is likely to resist possible failure.

The method presented several assumptions:

According to Qian, et al. (2003), calculation of the minimum FS condition required the inclination angle (ω) of the interwedge force to be equal to zero. The authors assumed the interwedge forces (E_P or E_A) were perpendicular to the interface of active and passive wedges and acted at a distance of H/3 above the base of the interface. In addition, tensile strengths of any overlying geosynthetic materials in the liner system were not considered in the following analysis method. The definitions of the parameters involved in the following analysis are listed at the beginning of this book.

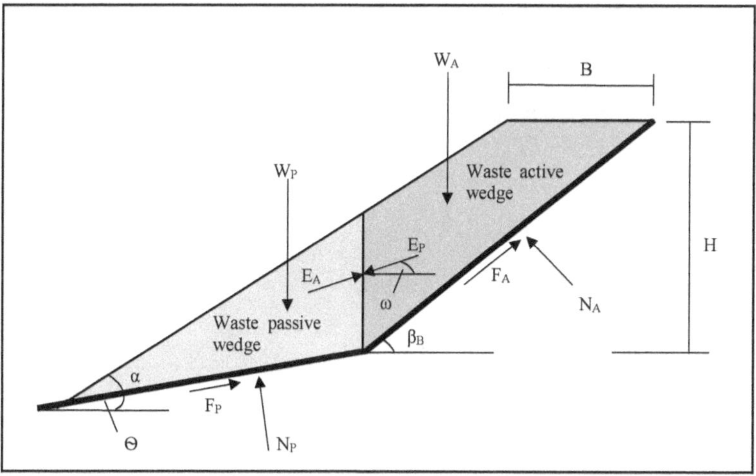

Figure 3-3: Two-part wedge analysis for a base lining system (Qian & Koerner, 2004)

Considering the force equilibrium of the passive and active wedges, the equations in Appendix II and Equation 3-1, the minimum factor of safety of the interface in Figure 3-3 was calculated.

$$FS = \frac{-b \pm \sqrt{b^2 - 4ac}}{2a}$$

Equation 3-1

Where,

$$a = (W_A cos\theta sin\beta_B) + (W_P cos\beta_B sin\theta)$$

$$b = \left(W_A tan\delta_p + W_P tan\delta_a\right) \times sin\beta_B \times sin\theta - \left(W_A tan\delta_a + W_P tan\delta_p\right) \times cos\beta_B \times cos\theta - C_A cos\theta - C_P cos\beta_B$$

$$c = -\left[\left(W_A cos\beta_B \times sin\theta + W_p sin\beta_B \times cos\theta\right) \times tan\delta_a \times tan\delta_p + C_A sin\theta \times tan\delta_p + C_p sin\beta_B \times tan\delta_a\right]$$

The values of C_A, C_{PA}, W_A and W_P in Equation 3-1 can be calculated using the following equations:

When B < H/tanβ_B,

$$C_A = c_a \times \frac{H}{sin\beta_B}$$

Equation 3-2

$$C_{PA} = c_p \times \left\{H - \left(H \times \frac{tan\alpha}{tan\beta_B}\right) + (B \times tan\alpha)\right\} / \left\{(cos\theta \times tan\alpha) - sin\theta\right\}$$

Equation 3-3

$$W_A = 0.5 \times \gamma_{SW} \times \left(\frac{H^2}{tan\beta_B}\right) - \left[0.5 \times \gamma_{SW} \times (\frac{H}{tan\beta_B} - B)^2 \times tan\alpha\right]$$

Equation 3-4

$$W_P = 0.5 \times \gamma_{SW} \left[\left(\frac{H}{tan\alpha} - \frac{H}{tan\beta_B} + B \right)^2 \times \frac{(tan\alpha \times tan\theta)}{(tan\alpha - tan\theta)} + \left[\left(\frac{H}{tan\alpha} - \frac{H}{tan\beta_B} + B \right)^2 \times tan\alpha \right] \right]$$

Equation 3-5

When B ≥ H/tanβ$_B$,

$$C_A = c_a \times \frac{H}{sin\beta_B}$$

Equation 3-6

$$C_{PA} = c_p \times \left(\frac{H}{tan\alpha} - \frac{H}{tan\beta_B} + B \right) \times \frac{\left[1 + \frac{tan\theta}{tan\alpha - tan\theta} \right]}{cos\theta}$$

Equation 3-7

$$W_A = 0.5 \times \gamma_{SW} \times \left(\frac{H^2}{tan\beta_B} \right)$$

Equation 3-8

$$W_P = \gamma_{SW} \left[0.5 \times \left(\frac{H}{tan\alpha} - \frac{H}{tan\beta_B} + B \right)^2 \times \frac{(tan\alpha \times tan\theta)}{(tan\alpha - tan\theta)} + H \left[\left(B - \frac{H}{tan\beta_B} \right) + 0.5 \times \frac{H}{tan\alpha} \right] \right]$$

Equation 3-9

3.5.1.2 Final capping system (Veneer liner)

Translational failure can occur as sliding of final cover systems along landfill capping slopes as seen in Figure 3-1. Similar to the base liner design method, a two-part wedge approach can be used to examine the stability of a final capping system. A finite length slope analysis shown in Figure 3-4 provides a reasonable representation of the potential failure surfaces in a capping system. The method is based on the computation of the balance of forces acting on a passive wedge at the toe and a long thin active wedge extending the length of the finite slope separated from the remaining cover soil by a tension crack at the crest. The cover soil is limited by an assumed critical failure surface between geosynthetic interfaces. Once again, the definitions of the parameters involved in the following analysis are listed at the beginning of this book.

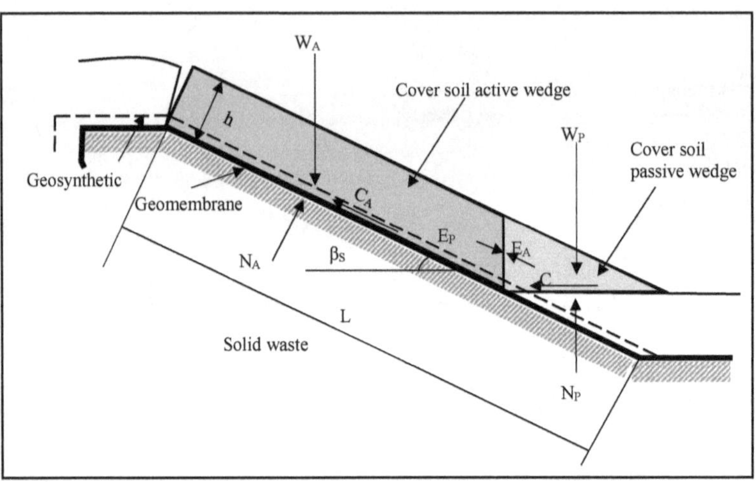

Figure 3-4: Two-part wedge limit equilibrium forces involved in a finite length slope analysis for a uniformly thick cover soil over multi-layered geosynthetics (Koerner & Soong, 1998)

The factor of safety for a cover lining system can be achieved by balancing the forces in the active and passive wedges as seen in Appendix II and using Equation 3-10. This analysis method is available from Giroud & Beech (1989), Koerner & Hwu (1991), Ling & Leshchinsky (1997), Koerner & Soong (1998) and Qian, et al. (2003).

$$FS = \frac{-b \pm \sqrt{b^2 - 4ac}}{2a}$$

Equation 3-10

Where,

$$a = (W_A - N_A cos\beta_S - T sin\beta_S)cos\beta_S$$

$$b = -[(W_A - N_A cos\beta_S - T sin\beta_S)sin\beta_S \times tan\varphi + (N_A tan\delta_c + C_a)sin\beta_S cos\beta_S + sin\beta(C + W_P tan\varphi)]$$

$$c = (N_A tan\delta_c + C_a)sin^2\beta_S \times tan\varphi$$

The values of C_a, C, N_A, N_P, W_A and W_P above can be calculated using the following equations:

$$W_A = \gamma h^2 \left(\frac{L}{h} - \frac{1}{sin\beta} - \frac{tan\beta_S}{2}\right)$$

Equation 3-11

$$N_A = W_A cos\beta_S$$

Equation 3-12

$$C_a = c_a \left(L - \frac{h}{sin\beta_S}\right)$$

Equation 3-13

$$W_P = \frac{\gamma h^2}{sin2\beta_S}$$

Equation 3-14

$$N_P = W_P + E_P sin\beta_S$$

Equation 3-15

$$C = \frac{(c)(h)}{sin\beta_S}$$

Equation 3-16

Where further reinforcement against instability is required and improvements of FS are essential geogrids can be used. The geogrids can provide an additional reinforcement force "T" that acts parallel to the slope or failure interface. This force "T" provides additional stability and acts only within the active wedge. The value of "T" in the design formulation is $T_{allowable}$, which is invariably less than the as manufactured tensile strength of the geosynthetic reinforcement material due to the reduction factors applied as illustrated in Equation 3-17.

$$T_{allowable} = T_{ultimate} \left(\frac{1}{RF_{ID} \times RF_{CR} \times RF_{CBD}} \right)$$

Equation 3-17

Where,

$T_{allowable}$ =   allowable value of reinforcement strength

$T_{ultimate}$ =   ultimate (as manufactured) value of reinforcement strength

RF_{ID} =   reduction factor for installation damage

RF_{CR} =   reduction factor for creep

RF_{CBD} =   reduction factor for long term chemical/ biological degradation

Another key element for stability calculations is the evaluation of settlement of all the waste layers and their intermediate soil covers, the soil and foundation materials beneath the landfill site, all the liner and leachate collection systems and all the final cover components (Shukla & Yin, 2006). When settlement/subsidence of waste layers is essentially complete, the previous cover should be replaced or can be incorporated into the final cover system (Shukla & Yin, 2006).

Stability analyses conducted by Kamon et al. (2003), identified the weakest interface is generally between woven geotextile component or composite clay liner and adjacent materials in landfill liners. As a result, waste containment facilities must consider the influence of geosynthetic material during slope stability analysis.

3.6 Factor of safety from slope stability data

Landfill slope designs must achieve factors of safety in respect of stability and integrity. A factor of safety is intended to account for uncertainty in design. The sources of uncertainty in the use of geosynthetics in landfill design could come from the following:

- Complexity of the ground conditions,
- Inadequacy of the information obtained from the site investigation,
- Variability of the materials,
- Laboratory testing (accuracy of test method),
- Effects of moisture on the barrier layer and
- Certainty/ accuracy of the design parameters e.g. shear strength and pore pressures.

Generally, the choice of an acceptable factor of safety value requires sound engineering judgement due to the multitude factors that must be considered. In South Africa, landfill slope designers tend to apply a minimum factor of safety (FS) of 1.3 as recommended by the Department of Water Affairs and Forestry (1998b). If the factor of safety is determined to be greater than or equal to the minimum FS, the slope is judged to be safe or to have acceptable stability. If the FS is less than 1.3, the slope is deemed unsafe.

4. Review of interface shear strength between geomembrane-geosynthetic combinations

4.1 Introduction

Since the implementation of geosynthetics in landfill lining systems, research has been undertaken to investigate their interface friction resistance between the various base liner or cover system components. Several researchers have conducted laboratory and field tests to determine the configurations that would provide optimal benefits (Bhatia & Kasturi, 1995; Triplett & Fox, 2001; Fox & Stark, 2015). From these studies different parameters were varied that included: type and stiffness of geosynthetics sheared; size and type of shear box, geosynthetic gripping devices and substrate used.

This chapter reviewed work carried out by previous authors that investigated friction behaviour of geomembrane/ geotextile, geomembrane/ GCL and geomembrane/ geogrid interfaces (collectively referred to as geomembrane/ geosynthetic interactions) commonly found in landfills. The chapter ends with conclusions drawn from gaps that could lead to improvements in research conducted in this field.

The dimensions of the direct shear test apparatus, the gripping and clamping devices, the test arrangements as well as the physical characteristics of the research materials investigated (e.g. geosynthetic mechanical properties, geomembrane surface roughness, geotextile and geogrid formulation, GCL reinforcement etc.) varied from researcher to researcher. Therefore, a direct comparison of the results in many cases was not possible in view of the unknown consistancy in tested materials and test apparatus used.

4.2 Direct shear test experimental approach

Pullout, ring shear and direct shear tests can be performed on geomembrane/ geosynthetic interfaces to characterise the shear strengths of geosynthetic combinations. From the three tests, the direct shear is the most widely used test to determine the maximum and ultimate coefficients of friction of geomembrane against geosynthetic interfaces (Bhatia & Kasturi, 1995; Bergado, et al., 1997; Bacas, et al., 2015).

The test apparatus has a primary function to determine the maximum shear stress at a geosynthetic/ geosynthetic interface for various normal pressures. This device is a modification of the conventional standard direct shear apparatus used to determine the angle of internal friction of soils. The modified test apparatus, illustrated in Figure 4-1 and Figure 4-2, is most suited to shear various types of geosynthetic products unlike the conventional standard direct shear device.

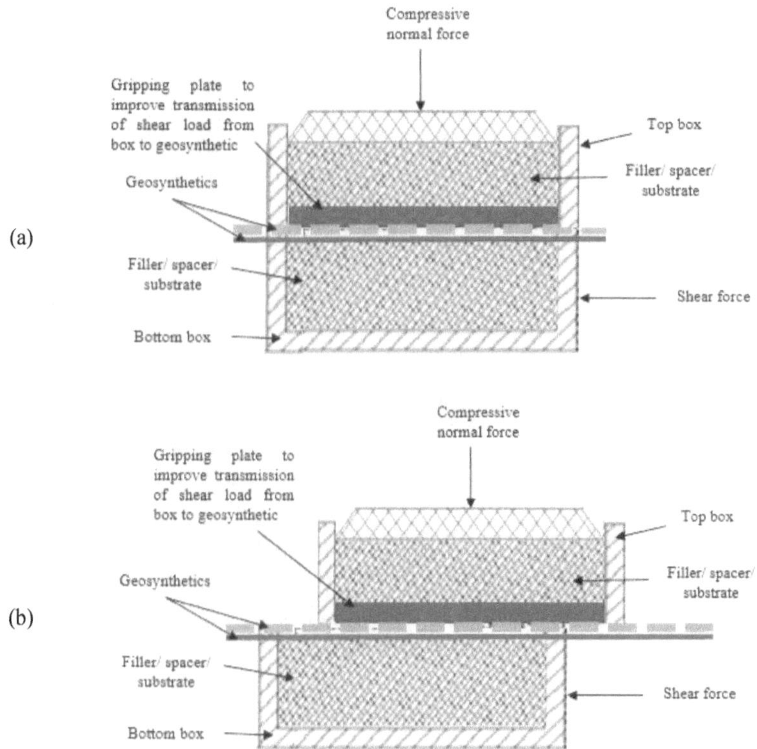

Figure 4-1: Examples of cross section of direct shear box with varying shear area arrangement (a) before and (b) after shear has taken place

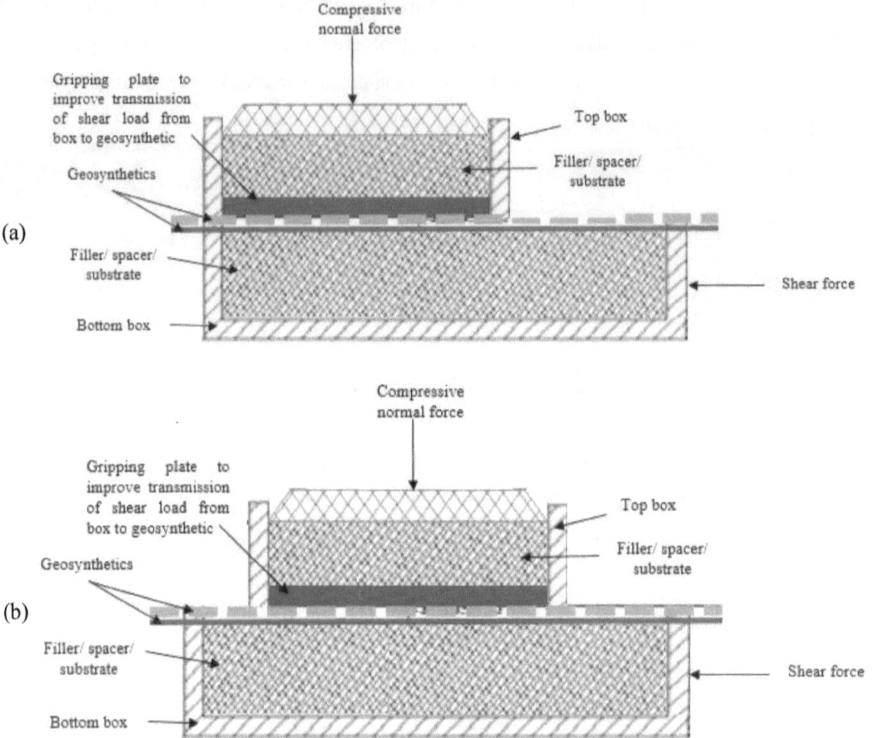

Figure 4-2: Examples of cross section of direct shear box with constant shear area arrangement (a) before and (b) after shear has taken place

In principle, there have been two direct shear test apparatus with different dimensions that have been adopted by researchers;

(a) A small square direct shear box; also referred to as the standard shear box has dimensions of 100 x 100 mm (Tuna & Altun, 2012; Padade & Mandal, 2012; Zaini, et al., 2012) and

(b) A large square direct shear box; that has dimesions of 300 x 300 mm or more (Bhatia & Kasturi, 1995; Russell, et al., 1998; Triplett & Fox, 2001). ASTM D6243 and ASTM D5321 recommend that geosynthetic interface tests be carried out in a 300 mm square or rectangular size shear apparatus. This shear box has an advantage over the first in that it can shear relatively large specimens with minimal boundary effects during testing which allows the determination of the interface residual shear strength and no generation of eccentric normal loads during shear (Gomez, et al., 2000).

For both dimensions, the test device consists of two halves of the direct shear apparatus with the geomembrane specimen sample usually clamped to the upper frame. There are two exceptions by Russell (1998) and Geofabrics (2001) where the geomembrane was attached to the lower frame in some of the tests conducted. The geosynthetic to be sheared with the geomembrane is placed in the lower frame so that the two materials face each other. A normal pressure (vertical load) is uniformly applied to the loading cover overlying the top geosynthetic. The upper frame is then sheared at a constant rate over the lower frame and the shear force required is incrementally measured. The test is repeated for at least two additional, yet different, normal pressures. The shear stress - displacement data is recorded and plotted for each of the three normal pressures to produce three curves. By plotting the peak and residual shear stress against the corresponding normal pressure, a Mohr-Coulomb failure envelope for the interface test series is produced, where interface friction angles and adhesion values can be obtained.

The standards highlight the average shear stress developed across the interface is calculated from the measured reaction force exerted on the respective shear frame. By assuming zero friction between the upper and lower frames and taking the contact shear area to be constant, the average shear stress, τ, is given as:

$$\tau = \frac{F}{A}$$

Equation 4-1

where,

F   shear force acting along the geomembrane/geosynthetic interface,

A   interface contact (shear) area.

The maximum shear stress, τ_p, is obtained by substituting the maximum interface shear force experienced, F_{max}, as well as the interface area, A, into Equation 4 1.

In Table 4-1, a summary of the relevant published work for geomembrane/ geosynthetic interface testing using the direct shear apparatuses is presented. The table clearly shows a relatively wide range of test arrangements, pressures and test rates which have been covered by researchers.

Table 4-1: Geosynthetic interface tests conducted by several authors

Author	Shear area dimension (mm)	Test rate (mm/min)	Pressure range (kPa)	Tested interfaces		Average peak friction parameters	
				Material 1	Material 2	Friction angle (°)	Adhesion (kPa)
(Infante, et al., 2016)	100 x 100	1	15.7-62.8	Sand	Geogrid	51.5	12.7
(Bacas, et al., 2015)	300 x 300	5	25-450	HDPE GM	Nonwoven geotextiles	23	0
(Liu, et al., 2009)	450 x 450	1	42-187	Sand	Geogrid	33.9	11.56
(Kamon, et al., 2008)	Bottom 350 x 600, top 250 x 500	1	100-300	HDPE, PVC GM	Nonwoven geotextile	21	3
(Fowmes, et al., 2008)	300 x 300	1	10-50	LLDPE, HDPE GM	Nonwoven geotextile	28.6	6
(Orebowale, 2006)	Bottom 300 x 400, top 300 x 300	3	10-200	LLDPE, HDPE GM	Nonwoven geotextile	33	3.8
(McCartney & Zornberg, 2003)	305 x 305	1	25-300	HDPE GM	Woven side of GCL	41.9	12.1
(McCartney, et al., 2002)	300 x300	0.1	7-310	HDPE, VLDPE, LLDPE GM	Woven side of GCL	25.6	-
(Geofabrics, 2001)	400 x 300	3	25-200	HDPE GM	Nonwoven geotextile	16.7	6.5
(Triplett & Fox, 2001)	406 x 1067	0.1	1-490	HDPE GM	Woven side of GCL	23.7	0

Author	Shear area dimension (mm)	Test rate (mm/min)	Pressure range (kPa)	Tested interfaces		Average peak friction parameters	
				Material 1	Material 2	Friction angle (°)	Adhesion (kPa)
(Russell, et al., 1998)	Top 305 x 305 bottom 305 x406	3	25-200	HDPE GM	Nonwoven geotextiles	28	3.6
(Bergado, et al., 1997)	300 x 300	8	150-400	HDPE GM	Nonwoven geotextile	7.4	0
(Bhatia & Kasturi, 1995)	305 x 305	1	15.31-200	PVC, LDPE and HDPE GM	Nonwoven geotextile	17.4	2.87

4.2.1 Different test arrangements

It is important to note the direct shear apparatus can have two different test arrangements: (1) varying shear area arrangement; that is the geomembrane/geosynthetic interface shear area changes as the test proceeds used by McCartney & Zornberg, 2003; Bergado, et al., 1997, and (2) constant shear area arrangement; the geomembrane/ geosynthetic interface shear area remains constant throughout the test similar to the devices used by Russell, et al. 1998; Triplett & Fox, 2001; Hillman & Stark, 2001 and Kamon, et al., 2008. The test arrangements shown in the cross-secton in Figure 4-1 and Figure 4-2: Examples of cross section of direct shear box with constant shear area arrangement (a) before and (b) after shear has taken place, are typical of the devices employed in the above research studies.

There are two possible sample set-ups within the direct shear apparatus. There is the (1) single interface shear tests, where one interface is sheared at a time and the (2) multi-interface shear tests where several interfaces are tested simultaneously therefore allowing failure to occur along the weakest interface.

Fox & Stark (2004) highlight that although multi-interface tests provide a better simulation of field conditions, the tests provide strength parameters only for the failure surface and not for the other interfaces. Thus no information is obtained on how close the other interfaces were to failure. This makes multi-interface tests more difficult to interpret failure than single-interface tests. The use of single interface tests provides design strength parameters with little uneasiness.

In some of the reviewed research work, the lower frame with the geomembrane attached remained stationary when the top frame was being sheared (e.g. Fox & Kim, 2008; Fox & Stark, 2004), yet in others, the upper frame remained stationary while the lower one was being displaced during the test (e.g. Bacas, et al., 2015; Hillman & Stark, 2001). Amongst all the reviewed cases, no comparative study had been undertaken on the effects of the difference in shear box size, geosynthetic placement and shear application has on the interface friction characteristics.

4.2.2 Use of gripping and clamping systems

ASTM D6243 and ASTM D5321 recommend that geosynthetic interface tests be carried out in a 300 mm square shear apparatus with the lower geosynthetic specimen placed over a rigid substrate in the lower frame while the top frame is occupied by another geosynthetic. The substrate can be a steel block or other rough media fitting closely inside the lower frame of equal height less the thickness of the geosynthetic sample material to be placed on the frame.

Large direct shear substrates can be covered with a coarsened surface or gripping system to secure the test specimen to the shearing frames. This assists in transfering shear stresses to the test specimen and isolating the effects of slippage occuring at an unintended interface. The effect of insufficient grippage can lead to inaccurate measurements of shear stress–displacement behaviour and strength mainly by the reduction of the peak (but not residual) shear strength and an increase in the displacement achieved at peak strength. The use of proper gripping surfaces during shear could reduce difficulties in test data interpretation, provide good drainage and increase the accuracy and reproducibility of test results (Fox & Stark, 2004; Allen & Fox, 2007; Fox & Kim, 2008).

Several authors have demonstrated success using a variety of different gripping surfaces such as those illustrated in Figure 4-3. Some examples of gripping surfaces that have been used are double-sided adhesive tape, nail plates molded in epoxy with a high density of short sharp nails, sharp 1-2 mm tall triangular pyramid teeth gripping plate, 'truss plates' (used for wood truss construction) or more advanced modified truss plate gripping surfaces created by the Geosynthetic Research Institute called the GRI-GCL4 (Triplett & Fox, 2001; Fox & Stark, 2004; Allen & Fox, 2007; Bacas, et al., 2015).

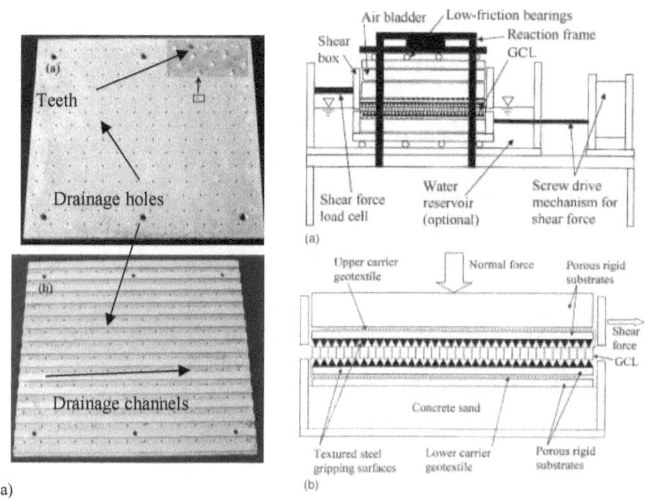

Figure 4-3: Textured gripping plates (a) Top side and bottom side (Bacas, et al., 2015) and (b) demonstrating how a steel gripping surface can fit into a direct shear device (Zornberg, et al., 2005)

An additional device to the gripping surface known as the clamping system could be used to increase the accuracy of results produced. This apparatus can consist of bolted bars (Figure 4-4), wrap-round mechanism (Figure 4-5(a)) or mechnical compression clamps (Figure 4-5(b)) that can fix the geosynthetics to one or both ends of the shearing frames during the shearing phase (Mackey & von Maubeuge, 2004). If a gripping surface is sufficiently rough, end clamping of the geosynthetics is not required (Fox, et al., 2014). The clamping procedures are able to perform a similar function to gripping systems, where they attempt to completely transfer shear stress through the outside surfaces into the geosynthetic and force failure to occur along the weakest interface (e.g., between the geosynthetic/

geosynthetic interface). If slippage occurs, tension may develop in the geosynthetics, which could lead to inaccurate shear strength results to be produced.

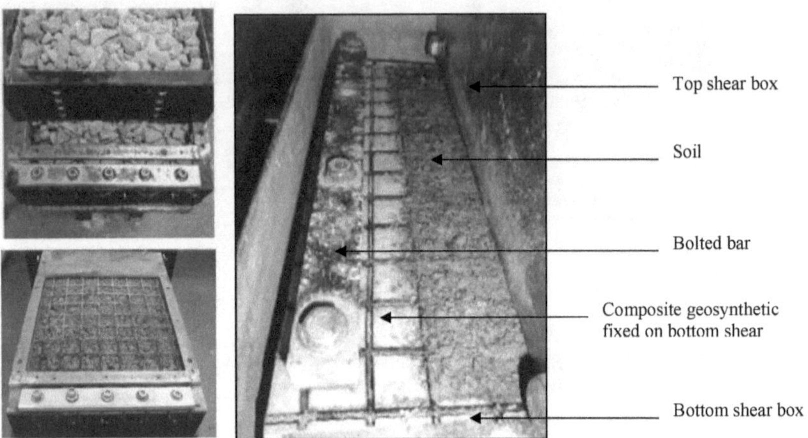

Figure 4-4: Bolted bar clamping device fixed at end of geocomposite/ soil interface (TRI Australasia, 2013; Arulrajah, et al., 2014a; Arulrajah, et al., 2014b)

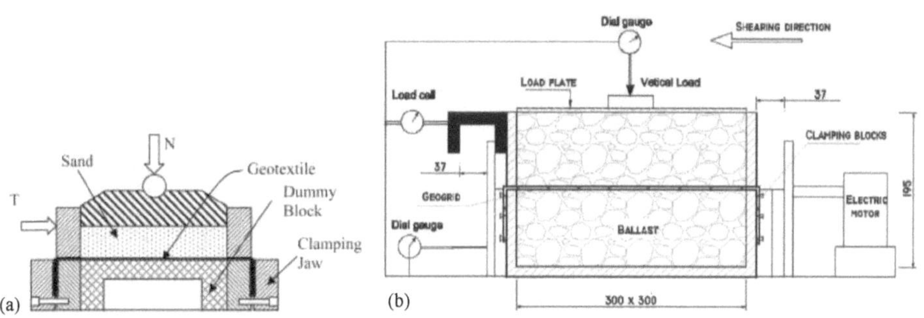

Figure 4-5: (a) Compression clamp (Anubhav & Basudhar, 2010) and (b) wrap-round mechanism (Indraratna, et al., 2011) fixing geosynthetics into place

Tensile failure of geosynthetics under shear is common in the field and in the laboratory. In the field, one of the ways tensile stresses can be induced in geosynthetics (usually those used in capping systems) is when localized failure occurs where underlying waste has been poorly compacted resulting in settlement of the waste and capping liner geosynthetics to be in tension (Shukla & Yin, 2006). In the laboratory, geosynthetics can experience tension near the clamping area when tested using direct shear box apparatus.

Several authors have shown that clamping mechanisms may result in the development of tension which could conclude with stretching, tearing or necking of the geosynthetic (Russell, et al., 1998; Bhatia & Kasturi, 1995; Fox & Kim, 2008; Fox & Stark, 2004). This indicates there is insufficient friction between the tested specimen and the gripping surfaces, which may introduce errors into the test data although the magnitude of the effect is currently unknown.

According to Fox & Kim (2008) the most accurate shear strength data will be obtained if the intended failure surface has the lowest shear resistance of all possible sliding surfaces. This means that the tested geosynthetics do not become tensioned at the clamps, shear displacement is nominally uniform at all points on the failure surface, peak shear strength occurs everywhere simultaneously and the measured relationship between shear stress and shear displacement is representative of true material behaviour.

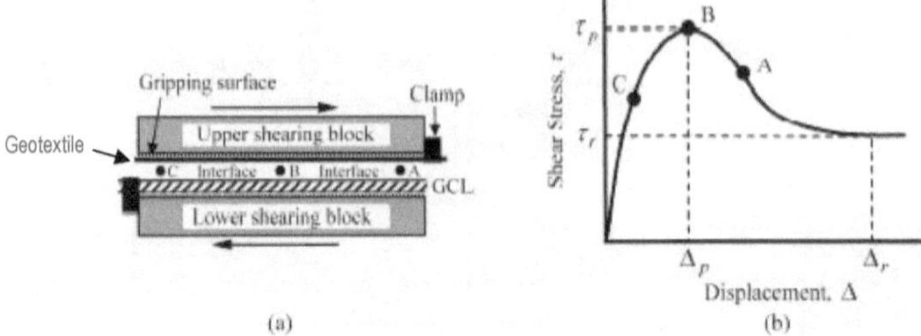

Figure 4-6: (a) Illustration of GMX/GCL interface shear test; (b) progressive failure of three interface points (Fox & Kim, 2008)

It can occur that the intended failure surface may not have the lowest shear resistance and failure may initially take place elsewhere. For example, assuming that the upper gripping surface and geotextile in Figure 4-6 have the lowest shear resistance, the geotextile will slide on the upper surface and become tensioned at the right clamp. Extension of the geotextile specimen causes shear failure along the geotextile/GCL interface to first occur at the right side (Point A) and then progress across the specimen to the left (Fox & Kim, 2008). Thus, shear displacement will not be uniform and different sections of the failure surface will have different shear stresses at any given time, therefore the measured shear stress – displacement relationship may not be a true representation of the material behaviour. Fox & Stark (2004) stated that the measured peak strength will be less than the actual peak strength for the interfaces but residual shear strength will remain unchanged. Russell, et al., (1998) supported Fox & Stark, (2004) by demonstrating that the residual interface shear strength is a property of the geomembrane and geotextile used and independent of the sustrate material used during testing.

In research conducted by Bhatia & Kasturi (1995), after flexible PVC geomembranes experienced initial failure of the interface, the continued increase in the shear stress caused failure with respect to the base (i.e. The entire interface slid over the compacted silty clay base material). Bhatia & Kasturi, (1995) believed that this facilitated the stretching of the sample and resulted in lack of peak failure stress in frictional strength of flexible interfaces. Considerable stretching of all polyvinylchloride (PVC) geomembranes sheared against HDPE geotextiles was observed.

Bhatia & Kasturi (1995) observed geomembranes stretching at high normal stresses (greater than 100 kPa). After reaching yield stress of the interface, PVC interfaces did not fail but maintained stability by stretching of the geomembrane material without loss of strength. This resulted in higher shear strength to be mobilised at large shear displacements, which was considered to be a function of stretching of the geosynthetic.

4.3 Mobilisation of shear stresses

Maximum shear stress obtained in direct shear tests occurs after a certain amount of horizontal displacement has taken place at the tested interface. This displacement to the peak shear stress is called shear stress mobilisation.

Researchers have been keen to investigate the shear stress development at the geomembrane/ geosynthetic interface during testing in anticipation of predicting the geomembranes' performance in containment facilities.

Orebowale (2006) illustrated shear stress mobilisation by conducting direct shear interface tests on smooth HDPE and LLDPE geomembranes sheared against nonwoven geotextiles. The results from their work was presented as shear strength - displacement plots for different geomembrane/ geotextile combinations. The plots indicated that maximum shear stress was obtained at different shear strengths and horizontal displacements for the various interfaces tested. The test results showed the interface shear strength decreasing with increasing shear displacement a phenomenon known as strain softening.

The mobilisation of the shear stresses occurred over relatively small shear displacements along both HDPE and LLDPE geomembranes when sheared over the nonwoven geotextiles. The shear stresses had a sharp peak between approximately 2 mm displacement, which gradually decreased to a residual stress level indicated by a relatively constant value. In addition, it was noted that LLDPE interfaces presented larger shear strengths when compared to HDPE interfaces for all combinations sheared.

Similarly, Bacas, et al. (2015) highlighted shear stress mobilisation by conducting direct shear interface tests on various HDPE geomembranes and nonwoven geotextiles. The results from their work were illustrated as friction shear strength - shear displacement plots for different geomembrane/ geotextile combinations which are presented in Figure 4-7. Where GMr1, GMr3, GT1, GT2 and GT3 represents geomembranes with textured surfaces smaller than 1mm, geomembranes with textured surfaces greater than 1mm, needle-punched monofilament geotextiles, needle-punched staple fibre geotextiles and thermally bonded monofilament geotextiles respectively.

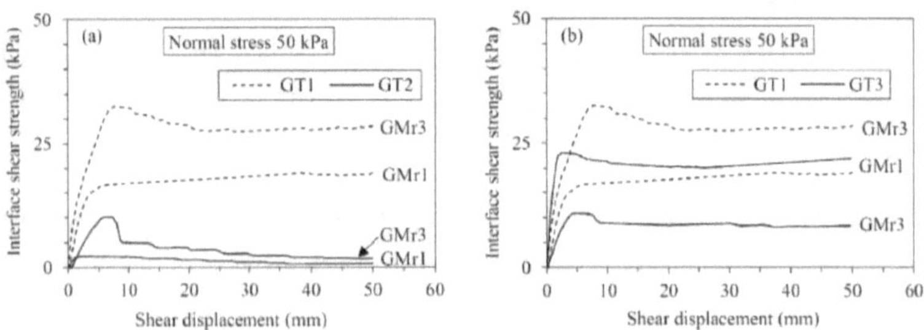

Figure 4-7: Illustration of shear stress mobilisation using various geomembranes and nonwoven geotextiles (Bacas, et al., 2015)

The plots in Figure 4-7 (conducted at an applied normal pressure of 50kPa) indicate that maximum shear stress was obtained at different shear strengths and horizontal displacements for the various interfaces tested. The test results show the interface shear strength decreasing with increasing shear displacment.

The mobilisation of the shear stresses occurred over relatively small shear displacements along both GMr1 and GMr3 when sheared over the three nonwoven geotextiles. The shear stresses had a sharp peak between approximately 2-7 mm displacement, which decreased to a residual stress level indicated by a relatively constant value as seen with GMr3/GT1, GMr3/GT2 and GMr3/GT3 interfaces. However for the geomembrane with a textured surface smaller than 1mm, i.e. GMr1, the plots showed gentle curves at peak strength with the shear stress increasing gradually over a relatively longer displacement to a not clearly defined residual stress value. This displayed that the shear stress at higher displacement was greater than at lower displacement. But this was not evident for the GMr1/GT3 interface. The interface appeared to have a relatively sharp peak at approximately 3mm which decreased to a residual stress level and then gradually increased to higher shear stress values from approximately 30 mm displacement. In addition, it was noted that GMr3 interfaces presented larger shear strengths for all combinations sheared except when tested against the GMr3/GT3 interface.

Similarly, Bhatia & Kasturi (1995) measured shear stress at 10 % strain along HDPE and PVC geomembranes sheared against nonwoven geotextiles. The results showed that HDPE geomembranes' shear stresses had a sharp peak which decreased to a residual stress level indicated by a constant value after the displacement of about 4 mm. The abrupt failure allowed for a clearly defined residual shear stress. The authors found the flexible PVC geomembranes experienced very large amounts of elongation during the direct shear tests before failure therefore the shear stress at 10 % strain indicated that the strength at higher strain was greater than at lower strain. It was evident from the plots that the PVC geomembranes did not show clear peak or residual shear stresses during shearing.

An investigation by McCartney, et al. (2002) showed that the mobilisation of the peak shear stresses occured over a relatively high shear displacement. The HDPE geomembrane/ GCL, Very Low Density Polyethylene (VLDPE) geomembrane/ GCL and LLDPE geomembrane/ GCL interfaces were found to have peak shear stresses at higher displacements namely ranging from 8 – 18 mm. It should be noted that the study was conducted using textured HDPE, VLDPE and LLDPE geomembranes sheared against hydrated needle punched GCLs which influenced the interface friction behviour.

Similar to Bhatia & Kasturi (1995), Hillman & Stark (2001) investigated textured flexible polyethylene (LDPE) and HDPE geomembrane tested against nonwoven geotextile interfaces which reached peak strength at displacements with a similar range of approximately 5-12 mm sheared over normal stresses 17-285kPa.

Bacas, et al. (2015) investigated coextruded HDPE geomembrane against non woven geotextile interfaces. It was concluded that peak shear stresses were reached at displacements between 4 and 10 mm over an applied normal stress range of 25-450 kPa. Although the geotextile/ geomembrane interfaces were tested under wet conditions, the water content did not significantly affect the interface shear strength.

In contrast to results found by Bacas, et al. (2015), Triplett & Fox (2001) observed a slower peak shear stress development for coextruded HDPE geomembrane sheared against non woven geotextile. The peak shear stresses occurred at a range of 12.2 - 21.3 mm shear displacement compared to the range of about 4 - 10 mm measured by Bacas, et al. (2015). This suggests that there are many variables influencing the mobilisation of shear stress.

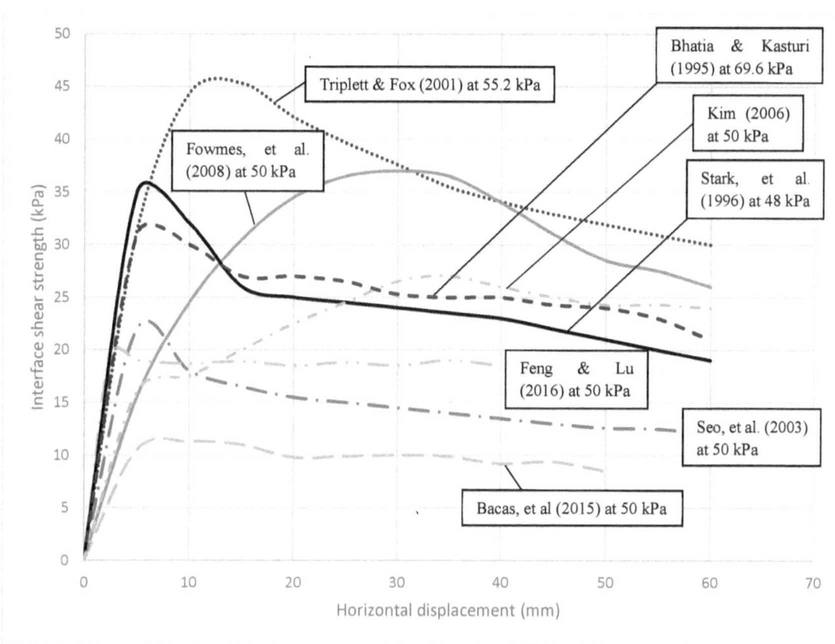

Figure 4-8: Comparison of shear displacements at peak from different literature

For example, Figure 4-8 illustrates a few literature results that demonstrate the difference in maximum shear strength and horizontal displacement at peak for a number of HDPE geomembrane/ nonwoven geotextile interfaces when sheared at similar confining stresses (range of 48 - 69.6kPa). The dissimilarity in mobilised shear stress can be related to factors that occur before and during shear. Among the causes of the variations in stress – displacement relationships include the differences in the shear device used, shear rate applied, physical properties of the geotextile, different geomembrane manufacturer, varying geomembrane thickness and asperities, unknown damage to geosynthetics prior to testing and many other aspects. Thus it is critical to test site specific materials before using shear strength parameters in design.

4.4 Interface shear stress-displacement relationship

The interface shear stress versus displacement relationship can be described using mathematical functions. An interface constitutive model that combines a non-linear hyperbolic model (Clough & Duncan, 1971) with a displacement-softening model (Esterhuizen, et al., 2001) was proposed to describe interface shear stiffness changes during shear (Gomez, et al., 2000; Wu, et al., 2011). The geosynthetic interface modelling can be performed by dividing the entire shear stress versus displacement graph into three regions; pre-peak region, strain softening region and residual region as shown in the Figure 4-9.

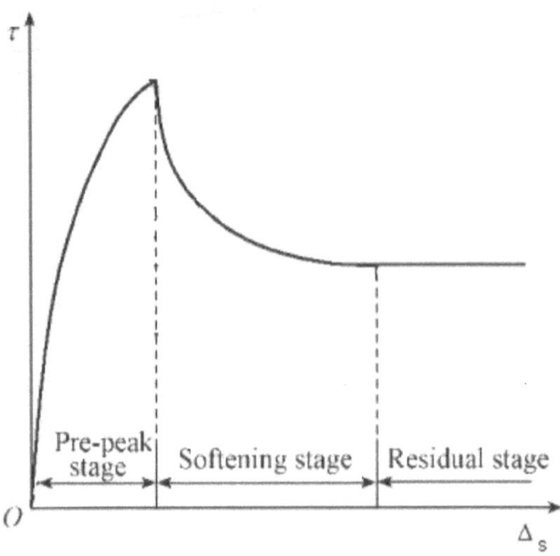

Figure 4-9: Shear stress – horizontal displacement relationship divided into three regions: pre-peak, strain softening and residual stages

4.4.1 Pre-peak

A non-linear hyperbolic model shape, in the pre-peak stage, can typically represent the relationship between shear stress and horizontal displacement at an interface (Wu, et al., 2011). The hyperbolic model for geosynthetic interfaces has been used extensively in design of geotechnical structures and often provides an accurate approximation to the interface response at constant normal stress (Gomez, et al., 2000). The procedures for identifying the stress-displacement relationship at geosynthetic interfaces are as follows:

(1) Figure 4-10 illustrates a reasonable approximation of a non-linear stress-displacement relationship on plotted data from an interface shear test. Figure 4-11 shows the same test data as Figure 4-10 plotted in terms of $\Delta s / \tau$ and Δs. If the interface shear stress-displacement behaviour follows a hyperbolic relationship, the plot in Figure 4-11 will be a straight line. The hyperbolic parameters 'a' and 'b' will be the vertical axis intercept and slope of this straight line respectively. If the actual interface test data does not exactly follow a hyperbolic relationship, the plot in Figure 4-11 must then be fitted to a straight line to determine the hyperbolic parameters 'a' and 'b'. The hyperbolic parameters 'a' and 'b' can be expressed as: $1/a$ = initial shear modulus of the interface (Ksi) and $1/b$ = asymptotic ultimate shear strength of the hyperbola (τ_{ult}). The value of τ_{ult} is usually assumed to be larger than the actual interface shear stress τ_f (or τ_p) (Gomez, et al., 2000; Seo, et al., 2003).

Clough & Duncan (1971) recommended that the best fit to the data can be obtained when the hyperbola intersected the test data at 70 and 95 percent of the strength shown in Figure 4-10 and Figure 4-11 as points P_1 and P_2 as well as points P_1t and P_2t respectively. A straight line is drawn between P_1t and P_2t allowing the hyperbolic parameters to be found as illustrated in the figures (Gomez, et al., 2000).

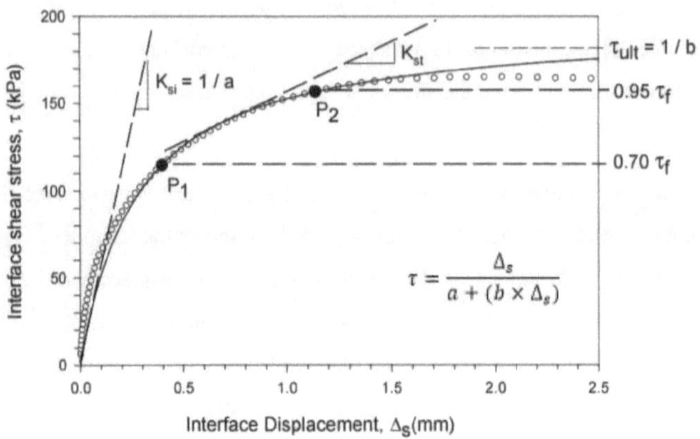

Figure 4-10: Plot of direct shear test data and the hyperbolic model (Gomez, et al., 2000)

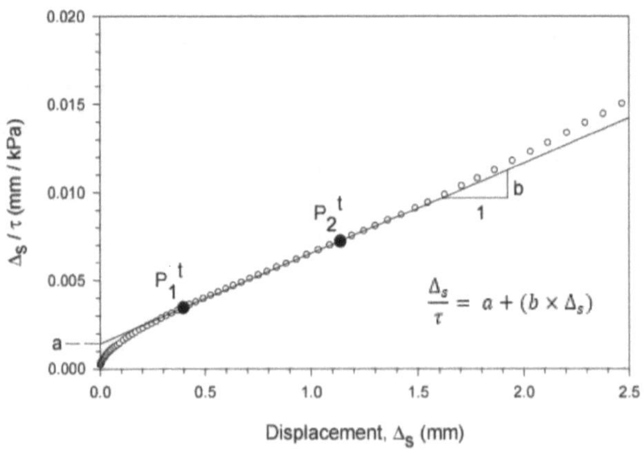

Figure 4-11: Determination of hyperbolic parameters a and b (Gomez, et al., 2000)

(2) Gomez (2000), Seo, et al. (2003) and Reddy, et al. (1996) indicated that the relationship between initial interface shear modulus and normal stress may be calculated using the following expression:

$$K_{si} = K\gamma_w \left(\frac{\sigma_n}{P_a}\right)^n$$

Equation 4-2

Where,

Ksi   =   the initial shear modulus (kN/m^3),

σ_n =   normal stress (kPa),

γ_w =   unit weight of water (kN/m^3),

K   =   dimensionless shear coefficient,

P_a =   atmospheric pressure (kPa), and

n   =   dimensionless modulus exponent.

(3) The shear coefficient (K) and modulus exponent (n) from Equation 4-2 can then be calculated by fitting Ksi/γ_w and σ_n/Pa values to logarithmic axes (Gomez, et al., 2000). The vertical axis intercept of the best-fit line gives the value of K and the gradient of the line gives the value of n as illustrated in Figure 4-12.

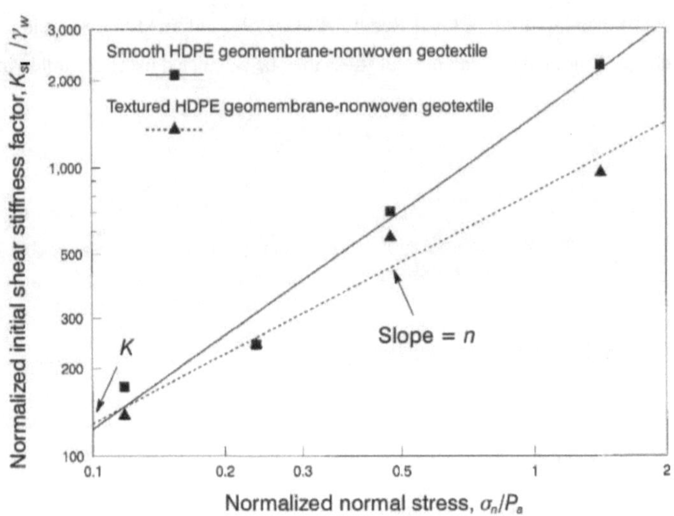

Figure 4-12: Determination of shear coefficient and modulus exponent (Reddy, et al., 1996)

(4) Duncan and Chang (1971) proposed that the tangent modulus (Kst) value at any point during shear could be obtained with Equation 4-3. Seo, et al (2003) noted that increasing value of n implies that the dependency of the Kst on normal stress increased at the pre-peak region. Therefore, sheared interfaces have the initial shear stiffness dependent on the magnitude of normal stress while the tangent shear modulus value is found to vary with the horizontal displacement in the tests.

$$K_{st} = \frac{\Delta \tau}{\Delta s} = K\gamma_w \left(\frac{\sigma_n}{P_a}\right)^n \left(1 - R_f \frac{\tau}{\tau_p}\right)^2$$

Equation 4-3

Where,

$$R_f = \frac{\tau_f}{\tau_{ult}}$$

Equation 4-4

Where,

Kst   =   tangent shear modulus (kN/m^3),

R$_f$ =   failure ratio,

τ_p or τ_f =   interface shear stress at peak or failure (kPa), and

τ_{ult} =   ultimate interface shear stress (kPa).

The proposed hyperbolic model for interfaces has some important limitations regarding its use; the hyperbolic formulation does not model displacement softening of the interface thus another model needs to be implemented for the rest of the shear stress versus displacement graph (Gomez, et al., 2000).

4.4.2 Strain softening

In the softening stage, the shear strength initially shows a sharp reduction with the shear displacement and then shows a gradual strength reduction. Esterhuizen et al. (2001) developed a non-linear displacement softening model to describe geosynthetic interface behaviour at post-peak region.

(1) Transformation of the initial stress vs. displacement curves into new curves that relate the strength degradation to the strain - softening shear displacement can be achieved using the shear strength strain-softening factor R calculated using Equation 4-5(Seo, et al., 2003; Wu, et, 2011).

$$R = \frac{\tau_p - \tau_{pr}}{\tau_p - \tau_r}$$

Where,

τ_{pr} = the post-peak shear strength,

τ_p = the peak shear strength, and

τ_r = the residual shear strength.

The parameters τ_p and τ_r can be calculated by the Mohr-Coulomb criterion using Equation 4-6 and Equation 4-7:

$$\tau_p = \sigma_n tan\varphi_p + C_p$$

Equation 4-6

$$\tau_r = \sigma_n tan\varphi_r + C_r$$

Equation 4-7

Where,

ϕ_p = the peak friction angle of the interface,

C_p = the peak cohesion of the interface,

ϕ_r = the residual friction angle of the interface, and

C_r = the residual cohesion of the interface.

(2) Once the strain softening factor is determined, the shear displacement ratio D is required. It can be calculated as follows:

$$D = u^p / u_r^p$$

Equation 4-8

Where,

u^p =   the plastic shear displacement, and

u_r^p =   the plastic shear displacement where the shear stress just reaches the residual strength.

Figure 4-13 provides a generalized shear stress versus displacement relationship and illustrations of parameters used in the strain softening modelling.

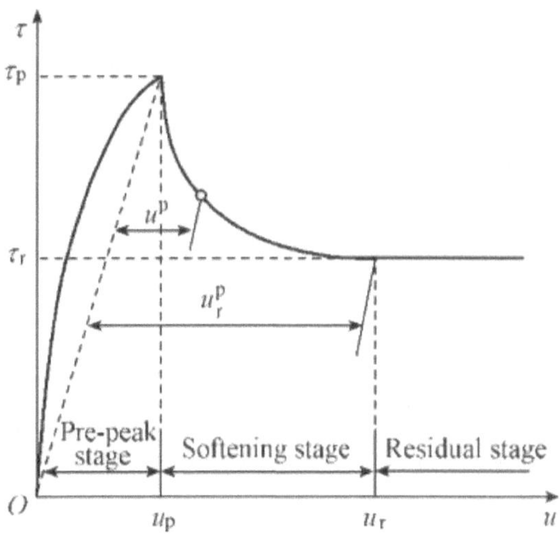

Figure 4-13: Shear stress and displacement relationship of geosynthetic interface (Wu, et al., 2011)

(3) Determination of the equation relating to R and D can be demonstrated by using Equation 4-9 and the knowledge that the curve passes through (1,1) (Seo, et al., 2003). The strength residual factor R and shear displacement ratio D curve can be plotted and approximated by a single hyperbolic relationship shown in Figure 4-14. From the equation K_0 values can be determined. Large values of K_0 would mean that interface shear strength decreased significantly at the early stage after peak strength was mobilized (Seo, et al., 2003).

$$R = \frac{K_0 D}{1 + (K_0 - 1)D}$$

Equation 4-9

Where,

K_0 =   initial slope of the R-D curve

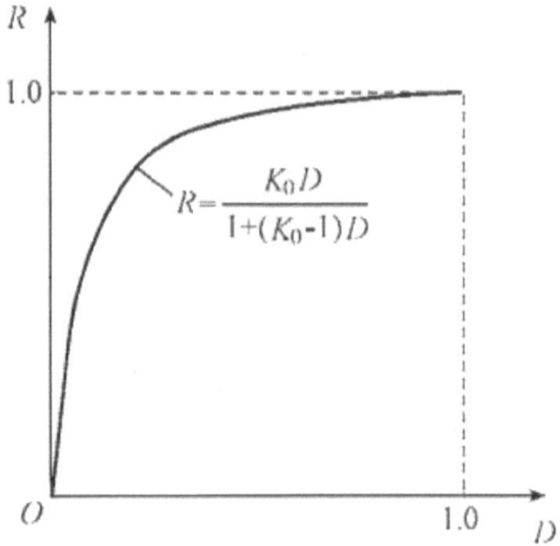

Figure 4-14: Typical R-D curve (Wu, et al., 2011)

After reaching the strain softening stage, the shear stress reaches a stable condition where it stays at a fairly constant value (τ_r) with the continuous increase of shear displacement known as the residual region (shown in Figure 4-13).

4.5 Shear stress versus normal pressure relationship

The shear stresses are obtained from the shear stress - displacement curves mentioned in Section 4.3 and Section 4.4 and are plotted against the normal pressure to represent a best fit line. This line can be linear, bilinear or non-linear and can give the maximum shear stress (τ_p) - normal pressure (O_n) relationship or residual shear stress (τ_r) – normal pressure (O_n) relationship both known as the Mohr-Coloumb failure envelopes (Russell, et al., 1998; Triplett & Fox, 2001; Fox & Stark, 2004; Bacas, et al., 2015). According to Coulomb's law, interface shear strength is obtained in terms of a friction angle and an apparent cohesion or adhesion. The interface friction angle is defined as the angle the failure envelope line makes with the horizontal axis. The intersection of the failure envelope with the vertical axis gives the interface cohesion or adhesion.

4.5.1 Adhesion

In this book the term adhesion was used to refer to interface shear strength characterization between two different materials (e.g. Geomembrane/ geogrid surfaces) instead of the conventional term cohesion, which was reserved for descriptions of interface shear strength between similar surfaces (e.g. Soil/ soil interfaces).

In soil mechanics adhesion or cohesion influenced by negative capillary pressure could be referred to as apparent adhesion/ cohesion. Under "undrained" shear strength conditions, the resistance of sheared surfaces to being pulled apart could be enhanced by the surface tension in surrounding pore water (Davison & Springman, 2000). For example, sand has no shear strength. Yet apparent cohesion in sand can be noticed when water is present, where apparent cohesion is typically higher than true cohesion as illustrated in Figure 4-15. The sand grains stick together due to negative pore pressure. This allows the sand to form slopes when wet (building sandcastles is an example) but will not stand when dry or saturated. In addition, research by Hungr & Morgenstern (1984) found that the Mohr-Coulomb strength envelope was dependent on moisture content, had stability strongly dependent on infiltration and had low reliability cohesion or adhesion values.

A common approach by the BS 6906, the draft European Standards and other publications (Russell, et al., 1998; Thiel, 2001; McCartney, et al., 2002; Fox & Stark, 2004; Blond & Elie, 2006; Kim, 2006; Qian, 2008a; Feng & Lu, 2016) suggested that the apparent adhesion or cohesion of shear strength envelopes should be ignored if positive. Thus, strength envelopes were assumed to pass through the coordinate origin.

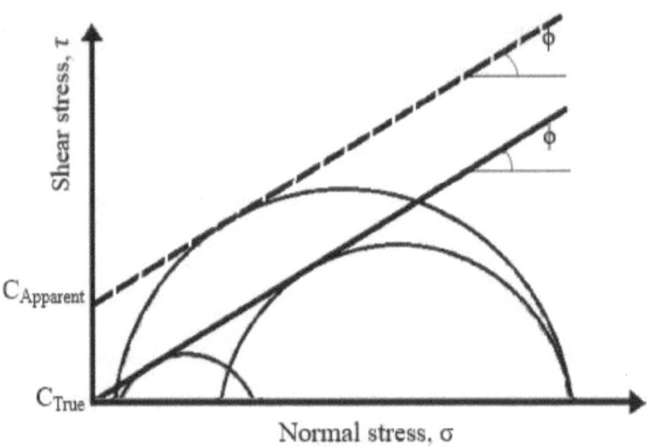

Figure 4-15: Mohr Coulomb failure envelope highlighting apparent adhesion

Failure envelopes that pass through the origin (i.e. have zero cohesion) were typical of GCL interface shear strengths of unreinforced GCLs (Fox & Stark, 2004). Several authors found that many failure envelopes of geotextile/ geomembrane interfaces and GCL/ geomembrane interfaces did not have large apparent adhesion components to the shear strength (Russell, et al., 1998; McCartney, et al., 2002; Feng & Lu, 2016). Some results reported negligible or even slightly negative values of apparent adhesion (Kim, 2006). This implied that interlocking between the geotextile/ geomembrane and GCL/

geomembrane interfaces was not present at low normal stresses (McCartney, et al., 2002; Feng & Lu, 2016).

It was common that apparent adhesion of geosynthetic surfaces was omitted in analysis methods due to the uncertainty in these parameter values. For products involving very high adhesions, ignoring the apparent adhesion of the materials had a notable effect in the shear strength of sheared materials as they would be quoted as being lower than measured. If a translational failure analysis method did not consider the apparent adhesion of the liner materials in a stability analysis, the mathematical analyses simplified and the interface between two materials that had the minimum friction angle for the multilayer liner system would be a critical potential failure plane with a minimum factor of safety (Qian, 2008a; Feng & Lu, 2016). Kim (2006) noted that simply ignoring the apparent adhesion seriously underestimated the value of the factor of safety. This resulted in significantly over conservative shear strength estimates for practical applications (Russell, et al., 1998; Thiel, 2001; McCartney, et al., 2002; Blond & Elie, 2006; Kim, 2006; Qian, 2008a).

Highly textured and reinforced geosynthetic materials usually have relatively high apparent adhesion values that allow for considerably improved shear strength stability (Qian, 2008a). Thus, in the authors' opinion the shear strength of most interfaces tested in this book presented important friction angles and apparent adhesion values. It was considered that both shear strength parameters (friction angles and apparent adhesion) would be measured and applied with great care in design purposes keeping in mind recommendations made by Russell, et al. (1998) that apparent adhesion should only be reported and used in an analysis within the range of normal stresses tested.

4.5.2 Interface friction angle

The geomembrane/ geosynthetic interface friction angle denoted in this disseratation by ϕ_A has been evaluated by several researchers (e.g. Bacas, et al., 2015; Fox & Kim, 2008; Triplett & Fox, 2001; Hillman & Stark, 2001) as indicated in Table 4-1.

According to Fox & Stark (2004) and Wasti & Ozduzgun (2001), interface friction angle values are influenced by the displacement rate, the surface roughness of the geomembrane and the raw material of the geomembrane (e.g. HDPE, PVC, LLDPE). In addition, Bacas, et al. (2011) found that friction angle increased due to the interlocking between the roughness of geomembranes and fibres of adjacent geosynthetics.

Bacas, et al. (2015) investigated the HDPE geomembrane/ geotextile interface friction parameters using a direct shear device. The researchers considered five textured geomembranes and one smooth geomembrane sheared against three nonwoven geotextiles. The interface friction angles of the smooth geomembrane combinations were found to be lower than those for geomembranes with higher surface roughness when sheared against all three nonwoven geotextiles. The different geomembrane/ geotextile interfaces tested presented peak and post-peak friction angle ranges demonstrated in Table 4-2. These results were confirmed by Fox & Stark (2004) and Russell, et al. (1998) who also obtained lower interface friction angles from smooth geomembrane/ geotextile interfaces when compared to textured geomembrane/ geotextile surfaces tested.

Table 4-2: Friction angles of NW PP staple fibre geotextiles sheared against different geomembrane interfaces (Extrapolated from Figure 7 in Bacas, et al., (2015))

Geomembrane asperity heights (mm)	Friction angle ($^\circ$)	
	Peak	Post-peak
0 (smooth)	8.1	5.7
0.25	19.8	11
0.50	23.8	11.6
0.80	25.1	12.9
1.20	24.6	9.5
1.30	29.5	12.4

Geomembranes with constant asperity heights but varying texturing types (e.g. coextruded and impingement) illustrated similar interface friction angles in an investigation conducted by Russell, et al. (1998). The coextruded textured geomembrane/ geotextile interfaces yield similarly high friction angles to the impingement textured geomembrane but with lower values of apparent adhesion as shown in Table

4-3. In addition, once again Table 4-3 demonstrated that lower friction angles were obtained for smooth geomembrane combinations when compared to textured geomembranes.

Table 4-3: Interface shear strengths of geomembrane/ geotextile combinations with different texturing styles (Russell, et al., 1998)

Geomembrane	Geotextile (g/m^2) (polymer)	Friction angle ($^\circ$)	Apparent adhesion (kPa)
Smooth	750 PP	8.4	3.2
	1200 PP	8.2	0.1
	800 HDPE	9.5	0.0
Impingement textured	750 PP	22.2	3.9
	1200 PP	20.7	3.2
	800 HDPE	20.4	5.2
Coextruded textured	750 PP	21.3	7.4
	1200 PP	22.1	5.8
	800 HDPE	21.0	7.7

In addition to varying texturing type, other geosynthetic manufacturing methods could also influence the friction angle obtained during direct shear tests. For example, research by McCartney, et al. (2004) demostrated that LLDPE geomembrane/ GCL combinations presented disimilar peak friction angles with changes in the GCL reinforcement. In the aurthors' research the peak friction angles obtained were (a) 20.6° for needle punched GCL (GCL B), (b) 26.3° for stitch bonded GCL (GCL C) and (c) 28.8° for thermally locked needle puched GCL (GCL A) all sheared against the same geomembrane. Figure 4-16 demonstrated the peak friction angles for the three different geomembrane/ GCL reinforcement types.

Furthermore, the raw material of a geomembrane have an effect on the interface friction angle values reached. Bhatia & Kasturi's (1995) study tested flexible Ethylene Propylene Diene Monomer (EPDM) geomembrane, medium stiff PVC geomemrbane and tough HDPE geomembrane using a large direct shear device. Their research showed that the interface friction angle of the the softer geomembranes was higher when compared to friction angles of the tougher geomembranes. Koerner (2005) and Williams &

Houlihan (1986) supported these findings. Williams & Houlihan (1986) found that for the same nonwoven geotextiles in contact with different geomembrane liners, the following interface friction angles were measured; Chlorosulfonated Polyethylene (Hypalone, 21°), Polyvinyl Chloride (PVC, 18°), Linear Low Density Polyethylene (LLDPE, 11°) and High Density Polyethylene (HDPE, 10°).

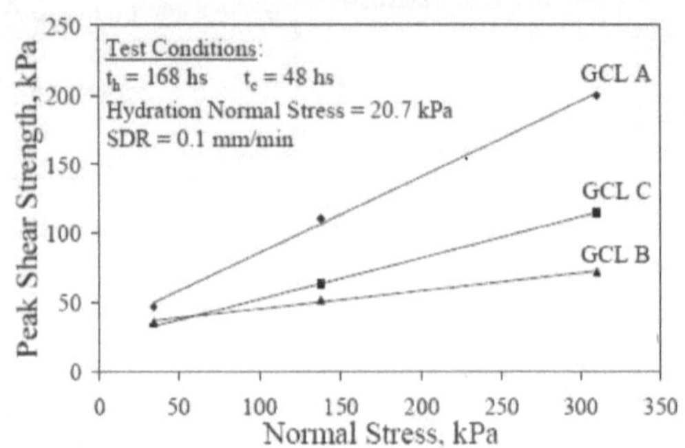

Figure 4-16: Variation of friction angle for GCLs with different reinforcement (McCartney, et al., 2004)

Hillman & Stark (2001) indicated that the peak and residual friction angles of textured HDPE, textured VFPE and faille PVC geomembrane/ GT2 nonwoven geotextile interfaces were 33°, 27° and 32° respectively and 15°, 19° and 26° respectively. It was noted; the textured VFPE geomembrane/ GT2 combinations achieved the least friction angle and underwent the smallest post-peak strength loss compared to the textured HDPE and faille PVC geomembrane interfaces. The reason for this small strength loss was that the flexible asperities of the textured VFPE geomembrane tore or pulled out a minimal amount of fibres from the geotextile, which allowed the geotextile to stay relatively intact and maintain interface strength.

An investigation of textured HDPE and PVC geomembranes sheared against nonwoven geotextiles by Kamon, et al. (2008) used direct shear tests (top box of 250 x 500 mm and bottom box of 350 x 600 mm) to examine the maximum shear stress (τ_p) versus normal pressure (O_n) relationship. The test device was of the type shown in Figure 4-2. The applied normal pressures ranged from 100kPa to 300kPa. The study indicated that all the maximum shear stress - normal pressure relationships were best-fit linear plots and varied with the type of geomembrane material. It was also clearly shown that wetting or submerging (geomembrane/geotextile interface) had no significant influence on the τ_p - O_n relationship slope.

Russell, et al. (1998) investigated the textured geomembrane/ geotextile interface friction behaviour using a specially designed direct shear apparatus (top box of 305 x 305 mm and bottom box of 305 x 406 mm). The tests were performed over an applied normal pressure range of 25kPa to 200kPa. The investigations were conducted under dry conditions unlike those of Kamon, et al. (2008). The τ_p - O_n relationship was plotted as a linear relationship for both HDPE and PP geomembranes tested against geotextiles. However, the authors noted that the relationship for the textured geomembrane/ geotextile interfaces would be described more accurately by non-linear failure envelopes.

Several researchers using the direct shear apparatus reported a non-linear τ_p - O_n relationship. They include; Bacas, et al., 2015 (investigating the interface behaviour between HDPE geomembrane and nonwoven geotextiles); Triplett & Fox, 2001 (considering HDPE geomembrane sheared against GCLs on both woven and nonwoven sides of cover geotextiles); Fox & Stark, 2004 (HDPE geomembrane and GCLs on woven and nonwoven sides of cover geotextiles); McCartney, et al., 2002 (hydrated needle-punched GCLs and textured LLDPE geomembrane); Fowmes, et al., 2008 (LLDPE geomembrane and nonwoven geotextile) and Hossain, et al., 2012 (geogrids against sandy, clayey and pure sand).

For instance, results obtained by Bacas, et al., (2015) using a shear apparatus (300 x 300 mm) of type shown in Figure 4-2, yielded non-linear τ_p - O_n relationships shown in Figure 4-17. Textured HDPE geomembranes were sheared against nonwoven geotextiles at applied normal pressures of 25kPa to 450kPa. Shearing was undertaken on hydrated geotextile/ geomembrane interfaces.

Although the straight envelopes show a good fit in Figure 4-17, it was clear that the non-linear trend curve (shown in a solid curve) illustrated a more accurate fit (Bacas, et al., 2015). The friction angle of the non-linear envelope was not constant throughout the range of the tested confining pressures. The angle of the slope of a tangent to this trend curve would decrease with increasing O_n (Fox & Stark, 2004; Bacas, et al., 2015). Interface test data from Bacas, et al. (2015) indicated that the peak interface friction angle of an investigated HDPE geomembrane and nonwoven geotextile combination decreased with increasing normal pressure from about 29° at low normal pressure (50kPa) to approximately 14° when the normal pressure was about 450kPa (refer to Figure 4-17).

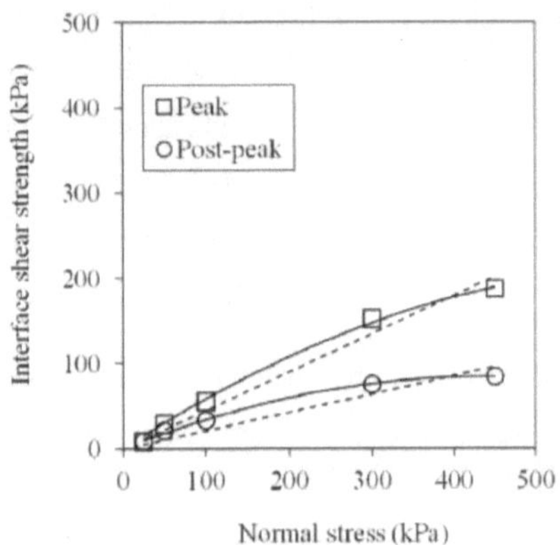

Figure 4-17: Peak and post-peak failure envelopes (linear and non-linear relationships) (Bacas, et al., 2015)

Fox & Stark (2004) found that peak and residual interfaces can be linear, multi-linear (e.g. bilinear, trilinear) or non-linear. Multi-linear interfaces gave abrupt changes in friction angle from approximately 100 kPa to 150 kPa normal stress (Triplett & Fox, 2001; McCartney, et al., 2004) while non-linear envelopes showed a gradual change in tangent friction angle as shearing normal stress increased (illustrated in Figure 4-18).

Fox & Stark (2004) also recognized that it was important to select and use the proper normal stress range for shear testing because a non-linear model appeared to be more appropriate over large stress ranges.

The normal stress range over which tests were conducted often dictated the degree of curvature in the resulting data. Geosynthetics in bottom liner landfill systems are subjected to a normal stress that is initially low and increases to a high value (as large as 1000 kPa or more) with time. Therefore, shear strength tests should be conducted for low, intermediate and high normal stress conditions in this case. On the other hand, geosynthetics placed in cover systems could be tested at low confining pressures due to these liners being subjected to low normal stress (approximately 10-25 kPa) that remains nearly constant after construction (Fox & Stark, 2004).

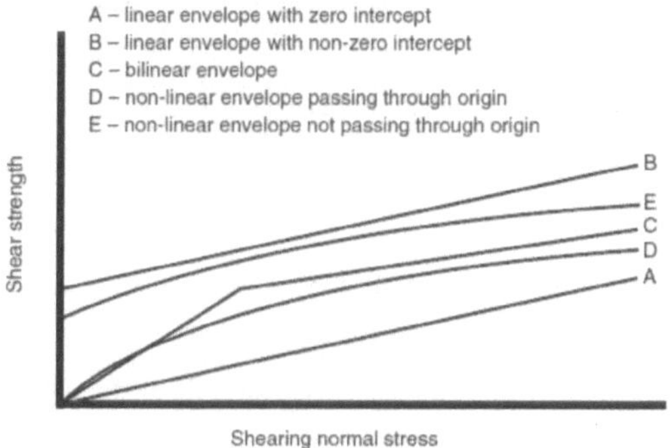

Figure 4-18: Examples of linear, bilinear and non-linear failure envelopes (Fox & Stark, 2004)

4.6 Effect of material structure on interface friction characteristics

Material structure has an influence on interface friction behaviour. This subsection reviewed geomembrane/ geosynthetic interaction mechanisms and how the geomembrane manufacture and formulation influenced shear strength charateristics.

4.6.1 Effect of geomembrane/ geosynthetic interaction mechanism

Generally, interface shear resistance increases as confining stress increases for most material interfaces. For example, the interaction mechanisms during shear tests on nonwoven geotextile/ textured geomembrane interfaces show the following behaviours:

(1) During the initial setting of the geotextile fibres on the geomembrane surface, the filaments of the geotextile tend to be attached on the geomembrane surface without external or internal force by a "velcro" effect (Geofabrics, 2001; Kim, 2006),

(2) At initial compression, low normal stresses (less than 50 kPa) allow the interaction to consist of friction between the superficial filaments of the geotextile and the asperities of the geomembrane (Bacas, et al., 2015). This interlocking bonding force can be described as the hook and loop effect and is quantified according to ASTM D 5169 standard. Hebeler, et al. (2005) found that the magnitude of hook and loop interaction was determined by the characteristics of the geomembrane texture and the tensile strain of the geotextile sheet,

(3) As the normal stress increases (greater than 50 kPa), the geotextile is compressed and the geomembrane asperities are introduced into the geotextile matrix, which is called interbedding (Bacas, et al., 2015). Thus, the high normal stress confines the geotextile dilation and result in high resistance and large displacement at peak due to the deep interlocking between the geotextile and geomembrane surfaces.

Several researchers have further investigated how interaction mechanism has led to peak interface shear strength for different interfaces (Stark, et al., 1996; Russell, et al., 1998; Geofabrics, 2001; Hillman & Stark, 2001; Triplett & Fox, 2001; Zornberg, et al., 2005; Kim, 2006; Shukla & Yin, 2006). According to Figure 4-19 by Kim (2006), in a textured geomembrane/ geotexitle combination (1) when normal stress is applied to the interface, geotextile density increases around geomembrane asperities. (2) As the shearing process begins, the geotextile fibres rearrange in an orientation parallel to the shear direction. This results in minor asperity deformation and high resistance to occur at relatively small shear displacement (peak displacement) during which the friction angle is first mobilized. (3) The resistance increases until the hook and loop mechanism takes place. At this stage, the geotextile begins to dilate with increase in shear displacement. (4) When the peak shear strength is reached, geotextile fibres slide over the geomembrane texture elements which causes the geotextile density to decrease at the interface and some geomembrane asperities begin to fail. (5) After the peak, the hook and loop mechanism degrades as the geotextile fibres that had embedded into the asperities of the textured geomembrane are pulled out, torn and untangled from the geotextile until the residual interface shear strength is reached. This causes large post-peak strength loss and the geomembrane texturing to be smoothened or polished during shear. The interlocking between the geotextile and geomembrane texture elements is influenced by the loose geotextile inner structure, heavily disturbed geotextile surface and deformed geomembrane texture elements.

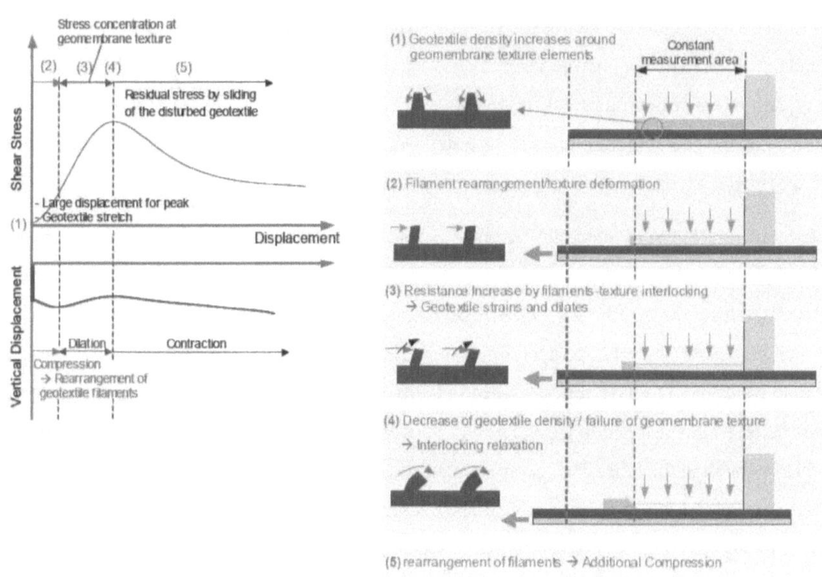

Figure 4-19: Geomembrane/ geotextile interaction mechanism (Kim, 2006)

Moreover in textured geomembrane/ geogrid interfaces, geogrids have been found to damage and smooth the asperities of the textured HDPE geomembranes during shearing thereby decreasing the roughness of the textured surface. This phenomenon was thought to reduce the shear strength achieved by the interface. It was noted that Hillman & Stark (2001) demonstrated that the surface of a faille PVC geomembrane can be roughened by a geogrid as shearing progressed.

Interestingly, Gilbert & Byrne (1996) found that the magnitude of post-peak strength reduction due to polished or smoothened geomembranes alone can be significant. In the authours' research, accumulated direct shear displacements were conducted on nonwoven continuous filament and needle-punched geotextile sheared along HDPE coextrued geomembranes. After each 25 mm increment of shear, the geotextile was replaced but the textured geomembrane specimen was not. Strain softening of these interfaces showed a maximum reductions from τ_p to τ_r of 62 % and 81 % for the nonwoven continuous filament and needle-punched geotextile respectively. These results indicate that geomembrane polishing has a significant strain-softening effect on nonwoven geotextile/ textured geomembrane interfaces. It must be noted that the geomembrane specimens tested were circular with a diameter of 60 mm, while the geotextile specimens were square with a length of 240 mm.

Hillman & Stark (2001) highlighted similar post-peak strength losses (50 % to 60 %) for a variety of textured HDPE geomembrane/ nonwoven geotextile interfaces. The researchers believed the reason for this large strength loss was that the asperities of the textured HDPE geomembrane tore or pulled out the fibres of the geotextile and the geomembrane texturing was smoothened or polished. The PVC geomembrane were found to have torn or pulled out only a small quantity of fibres from the geotextile. The authors believed this resulted in a trend of non-noticeable post-peak strength loss at normal stresses less than 50 kPa for all of the faille PVC geomembrane/ nonwoven geotextile interfaces tested. At normal stresses between 96 and 285 kPa, the residual shear strength of the PVC geomembrane interface was only approximately 15 % to 25 % lower than its peak shear strength. In addition, textured VFPE geomembrane interfaces were also investigated and experienced a larger post-peak strength loss than the PVC geomembrane interfaces. This was mainly to be because the VFPE geomembrane texturing tore or pulled out more fibres from the nonwoven geotextile during shearing than PVC geomembrane asperities. No indication of geomembrane elongation after testing was reported.

4.6.2 Effect of geomembrane composition

Hillman & Stark (2001) investigated the effect of geomembrane composition on interface friction parameters by looking into various geomembranes being sheared against a nonwoven PET geotextile (GT2). Figure 4-20 presents the shear stress-displacement results for faille PVC, textured HDPE and textured VFPE geomembrane/ GT2 geotextile interfaces at a normal stress of 192 kPa. From the figure, it can be seen that all three geomembrane/ geotextile interfaces appear to yield similar peak shear strengths and faille PVC combinations achieved considerably higher residual shear strengths than textured HDPE and textured VFPE geomembrane/ nonwoven geotextile interfaces. It was noted that the VFPE and HDPE geomembrane interfaces reached a peak strength condition after approximately 5 mm of shear displacement and then experienced a substantial post-peak strength loss (40 to 60%). On the other hand, the faille PVC interface peaked at a shear displacement of approximately 18 mm and lost only 20 to 25% of its peak shear strength (Hillman & Stark, 2001).

Bhatia & Kasturi (1995) conducted experiments on textured HDPE and PVC geomembranes. In the case of HDPE geomembranes, similar to findings by Hillman & Stark (2001), the stress-displacement response of the interfaces decreased to a stable stress value after reaching peak stress. Hence, the shear stress at 10% strain for the rigid geomembranes (HDPE) was less than at the peak. However for the flexible PVC geomembranes, due to stretching during the tests, the strength gradually increased with further shear. Thus, shear stress achieved at higher strain was greater than at lower strain. This was observed with all PVC interfaces when sheared against all the other interface materials. The researchers believed that under field conditions, if the PVC geomembranes were to be stressed beyond the yield stress for the interface, the material would stretch under the load without any loss of strength or material damage.

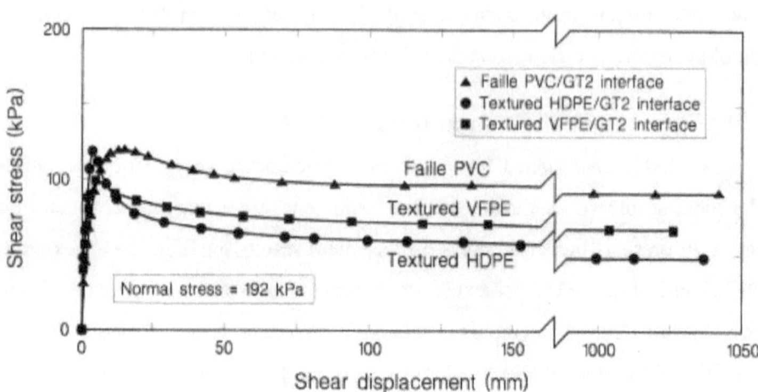

Figure 4-20: Comparison of shear stress-displacement relationships for faille PVC, textured HDPE and textured VFPE geomembrane/ GT2 geotextile interfaces (Hillman & Stark, 2001)

Kamon, et al. (2008) found that PVC geomembrane/ geotextile combinations produced higher friction angles than HDPE interfaces. Similarly, Martin et al. (cited in Bhatia & Kasturi 1995) conducted experiments on PVC, Chloro-Sulphonated Poly-Ethylene (CSPE), Ethylene Propylene Diene Monomer (EPDM) and HDPE geomembranes tested against sand, clay and geotextiles using a 100 mm x 100 mm shear box. Based on the results of the research tests, the authors concluded that the more flexible the geomembrane, the higher the friction angle obtained.

4.7 Summary of previous research

4.7.1 Key points

Possible shear failure of geotechinical engineering projects with geosynthetics incorporated has influenced designers and researchers to conduct laboratory tests on these materials. Investigation of the friction behaviour of geosynthetic contact surfaces allows for further understanding of the shear stress - horizontal displacement curve, shear stress – normal stress relationship, material type and apparatus used.

The review of past studies demonstrates that shear stress mobilisation occurs after a certain amount of horizontal displacement has taken place at a tested interface. This mobilisation forms the shear stress - displacement relationship which is a function of the particular geosynthetic/ geosynthetic combination.

In current practice, HDPE geomembranes often compete with LLDPE geomembranes as landfill capping liners. This literature review demonstrates that research involving LLDPE geomembranes is not adequately represented compared to that involving HDPE geomembranes, yet LLDPE geomembrane interface friction is critically important. Testing programs involving comparison of flexible (PVC, VFPE etc) and HDPE geomembranes have shown that stiff HDPE geomembranes have steeper shear stress - displacement graphs. There was evidence that flexible geomembranes tend to delay the appearance of peak shear stress at the interface tested and the residual shear stress did not vary significantly from peak shear for these geomembranes.

The research reviewed suggested that the maximum shear stress to normal pressure curve generated the Mohr-Couloumb failure envelopes which varied depending on the material involved. Literature showed that the strength envelopes with low correlation coefficients for linear curves could be better illustrated as curved or multi-linear failure envelopes. The failure model became non-linear when projected over a wide range of normal pressures. It was therefore important to simulate site specific loading conditions when running laboratory tests.

4.7.2 Gaps in knowledge

There is no one single geomembrane which is suitable for every containment application, the need to investigate if similar frictional behaviour patterns exist between HDPE and LLDPE geomembrane combinations was desired.

Testing programs involving comparison of flexible (PVC, VFPE etc) and HDPE geomembranes have shown that stiff HDPE geomembranes have steeper shear stress - displacement graphs. Questions on whether this behaviour was true for LLDPE geomembrane interfaces have been raised and prompted an investigation. Thus, the results of this study were anticipated to have a valuable contribution towards the selection of particular materials (between HDPE and LLDPE geomembranes) used in industry for landfill cover use.

Only a few experimental studies investigated at which critical confining stress the linear failure envelope became non-linear in a multi-linear failure envelope. This encouraged further research to better understand the mechanical behaviour of the geomembrane/ geosynthetic composites.

The spefic purpose of this research project is to address these current gaps in knowledge and provide possible new insight.

5. Research materials and methodology

5.1 Introduction

This chapter gives details of the testing equipment and methodology followed to determine shear strength of various geomembrane/geosynthetic interfaces. In this study, a modified large direct shear box test apparatus was chosen and used because of its ability to test various types of geosynthetic products. Three different types of geosynthetics commonly used in landfill capping and base liner systems were investigated namely; geotextiles, geosynthetic clay liners and geogrids. These geosynthetics were sheared against High Density Polyethylene (HDPE) and Linear Low Density Polyethylene (LLDPE) geomembranes. Finally, this chapter illustrates the data analysis methods used to obtain the design values under investigation.

5.2 Research materials

The following materials were used for the laboratory investigation:

5.2.1 Geosynthetic Clay Liners

Two needle punched geosynthetic clay liners (GCLs) were selected for investigation; namely GCLA and GCLB, frequently implemented in the base liner or capping system of a landfill as shown in Figure 3-9. These needle punched GCLs were chosen because they represented the most common reinforced GCL type used in South African landfills (GIGSA, 2009; Kaytech Engineered Fabrics, 2010a; Tancott, 2013a; E-Square engineering, 2015). Both GCLs had two polypropylene geotextiles (GT), one woven (W) and another nonwoven (NW) which formed top and bottom cover layers that sandwiched a layer of sodium bentonite powder (Figure 5-1). The nonwoven and woven geotextiles used in the GCLs achieved different purposes as illustrated in Table 5-1.

In order to keep the components together, fibre reinforcements were punched through the top cover geotextile through the sodium bentonite and the lower carrier geotextile forming a needle punched GCL (Kaytech Engineered Fabrics, 2014a). Entangled reinforcement fibres left on the surface to the geotextiles after the needle punching process were modified using a proprietary heat treating process thus ensuring that the fibres did not pull out from the geotextiles during sodium bentonite swell or during shearing. Permanently locking the fibres into place allowed for high shear strengths to be transmitted as tensile forces through the internal fibre reinforcements (Kaytech Engineered Fabrics, 2014a).

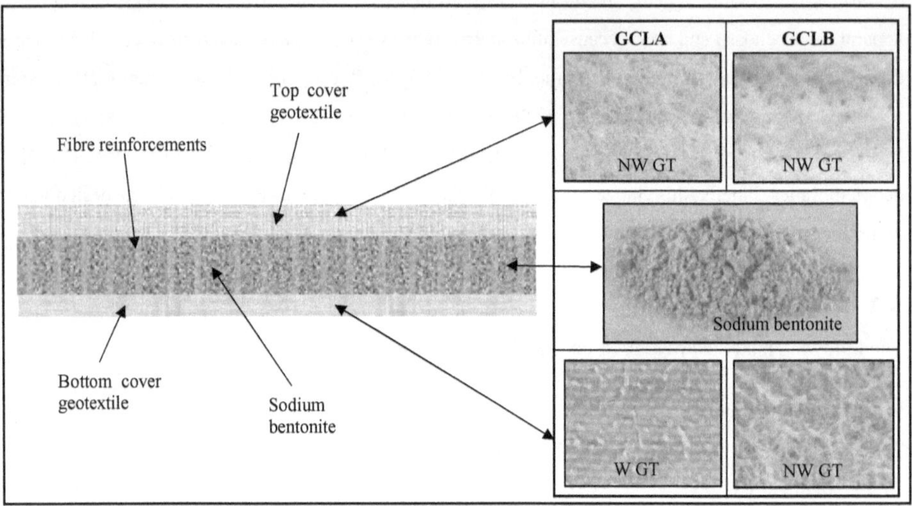

Figure 5-1: Components that make up a geosynthetic clay liner geocomposite

The GCLB was manufactured with an additional NW composite geotextile carrier layer illustrated in Table 5-2, unlike the GCLA. This composite layer enabled an increase in grab strength in both machine and cross machine direction as well as improved the hydraulic conductivity of the GCLB liner from 2.56 x 10^{-11} m/s of the GCLA to 1.92 x 10^{-11} m/s (Kaytech Engineered Fabrics, 2016). In practical design, the selection of GCLA, instead of GCLB, would provide a sufficient barrier that prevents contaminated water produced by low risk waste mainly in Class C and Class D lining systems from seeping into the surrounding natural environment. If there was a higher risk of contamination (i.e. Class A and Class B), a GCL with a reduced hydraulic conductivity would be required such as that provided by GCLB.

Table 5-1: Function of nonwoven and woven geotextiles in geosynthetic clay liner geocomposites

Nonwoven carrier geotextile	Woven carrier geotextile
Provides interlocking capabilities with other exterior interfaces	Allows filtration
Provides interlocking capabilities with the sodium bentonite clay	Provide tensile resistance to the GCL
Allows for in-plane drainage and filtration	Provides interlocking capabilities (although limited)
Provides puncture protection to the bentonite layer of the GCL	Allows easier bentonite extrusion when compared to nonwoven carrier geotextiles
Provides good connection for fibre reinforcements	

The bentonite sandwiched between the two geotextile layers had the following molecular formula,

$$Al_2H_2Na_2O_{13}Si_4$$

The sodium bentonite clay mostly consisted of the mineral, sodium montmorillonite. The bentonite quality of GCLA and GCLB was montmorillonite greater than 75 % and sodium cation (Na^+) greater than 60 % (Kaytech Engineered Fabrics, 2010b). According to the manufacturers of the GCLs, the quality of bentonite was controlled by purchasing from an ISO 9001:2008 accredited supplier (Kaytech Engineered Fabrics, 2014a).

Water would be able to cause the individual crystals of montmorillonite clay to expand, absorbing as much as several times its dry mass in water. The great increase in volume made the bentonite a useful sealant that would provide low permeability barriers. The swelling properties observed with hydrating sodium bentonite also made it very plastic and resistant to fracturing or cracking. Interestingly, bentonite could be hydrated and dried, frozen and thawed repeatedly without losing its original swelling capacity (CETCO, 2013). It was a combination of these physical properties that made sodium bentonite an ideal waterproofing material.

5.2.2 Geotextiles

The experimental program was conducted with two geotextiles commonly used in South Africa. The first geotextile (GTA) was selected because it had the standard inbuilt flexibility required to provide puncture protection to geomembrane liners in landfill systems (Geofabrics, 2016). The popular implementation of the second geotextile (GTB) has mainly been due to environmentally sustainable advantages (such as reduction in waste and use of raw materials, decrease in emission of greenhouse gases and lessening landfill space use) related to the geotextile being manufactured from 100 % polyester (Tancott, 2013b).

GTA was a nonwoven, polyester staple fibre needle punched geotextile (Figure 5-2(a)) with a thickness of 7.5 mm under 2 kPa. This geotextile had a permeability of 2.6 x 10^{-3} m/s at 50 mm head (Geofabrics, 2015). GTA can reduce puncture risks of the geomembrane during installation (Geofabrics, 2016). In a given design situation, these staple fibre geotextiles are installed prior to the placement of the aggregate layer to prevent damage to impermeable liners designed to contain harmful leachates in landfill applications as illustrated in Figure 2-14.

GTB was a nonwoven needle punched continuous filament polyester geotextile (Figure 5-2(b)). Being manufactured differently from GTA allowed GTB to have a thickness of 4.4 mm under 2 kPa and permeability 4 x 10^{-3} m/s at 50 mm head. In practical design, unlike GTA, the selection of GTB can be able to perform two functions in landfill applications; that of filter layer between the primary waste and the leachate collection system. GTB in-plane drainage characteristic could dissipate pore water pressure build-up beneath the liner while providing excellent filtration characteristics (Kaytech Engineered Fabrics, 2015). Similar to GTA, GTB provided resistance from piercing making this material ideal for cushioning protection for geomembranes.

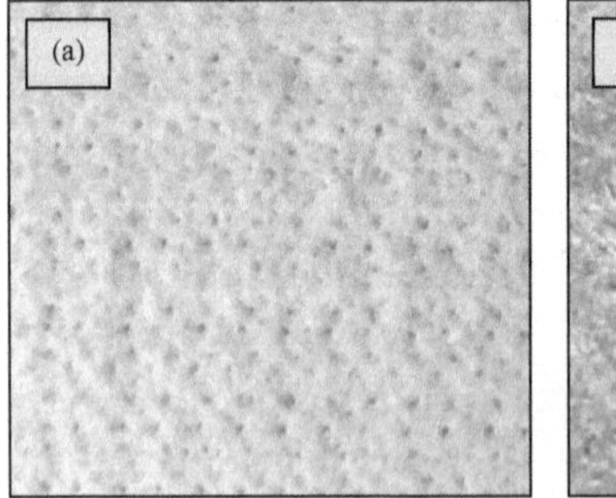
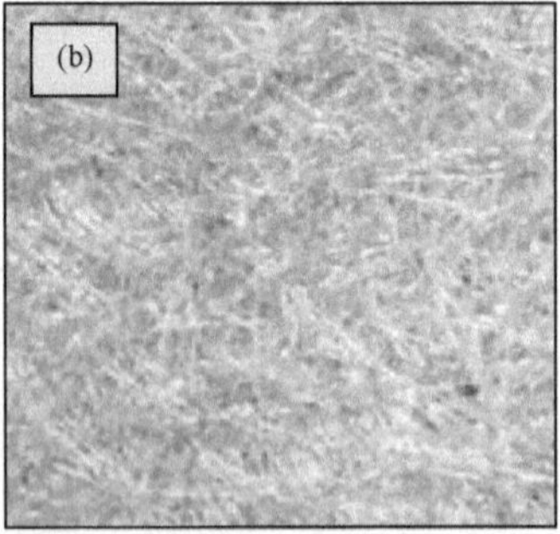

Figure 5-2: Nonwoven needle punched polyester geotextiles (a) Staple fibre – GTA and (b) Continuous filament – GTB

5.2.3 Geogrid

One geogrid suitable for reinforcement applications was selected to be used. This geogrid was chosen because of its regular use as reinforcement between geocells filled with soil and geomembranes (Figure 2-14) in South African landfill cover applications (Tencate, 2013).

When geocells are used in landfill cover systems, they provide lateral confinement and erosion control of the aggregate infill cover material (Foye, 2011). The self weight of the aggregate filled geocell (and any overburden, i.e. Snow) produces driving forces that induce sliding. Forces resisting sliding are interface friction between the contact surfaces. But friction alone can be insufficient to prevent sliding of the cover system, thus the cover requires additional support from tension reinforcement members anchored at the crest of slope. A geogrid reinforcement technique beneath the geocell layer can be applied, which allows all structural loads to be transmitted directly from the cover material to the geogrid. Engineering of the geogrid primarily concerns the selection of a geogrid with sufficient long-term strength to prevent sliding.

The selected geogrid was manufactured from black polymer coated high tenacity polyester (PET) yarns that were knitted to form a structured grid with polymeric coating protection (Figure 5-3). The geogrid's structure enabled it to provide high tensile strength, high fibre interlock strength, high soil interaction and pull out strength. The material had a short term tensile strength of 130 kN/m and 30 kN/m in the machine and cross machine direction respectively as depicted in Table 5-2: Summary properties of geosynthetics sheared against geomembranes.

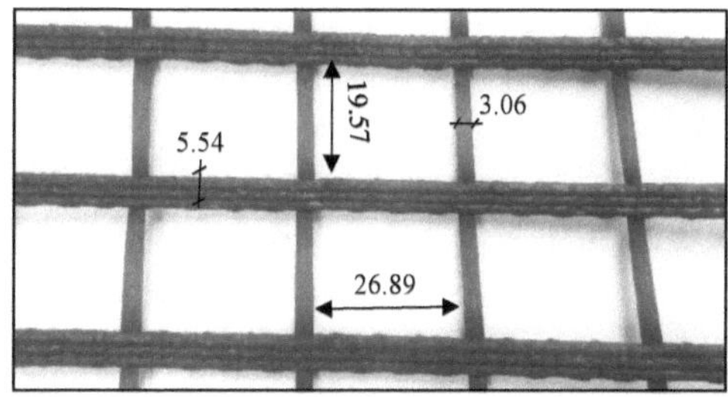

Figure 5-3: The geogrid tested with detailed dimensions (all dimensions in mm)

Table 5-2: Summary properties of geosynthetics sheared against geomembranes

	Units	Standard	GTA[1]	GTB[2]	GCLA[3]		GCLB[3]			Geogrid[4]
Material			PET	PET	Cover layer	Carrier layer	Cover layer	Carrier layer		PET
					PP	PP	PP	PP	PP	
Mass per unit area	g/m^2	ASTM D5261			NW	W	NW	W	NW	
					200	110	200	110	310	
Fibre type			Staple	Continuous	Staple	Slit film	Staple	Slit film	Staple	Coated yarns
Pore size	μm	ISO 12956:10	< 70	114						
Tensile strength	kN/m	ISO 10319:08	86	40						130/30
Elongation (200 mm wide strip)	%	ISO 10319:08	> 65	50-70	≥ 15		≥ 50			
Static puncture resistance (CBR)	kN	ISO 12236:06	14	7.1	1.4		2.5			
Puncture resistance (Max diameter of hole)	mm	ISO 13433:06	64	8						
Tear resistance	N	ASTM D4533	1500	1200						
Peel strength	N/m	ASTM D6496			> 360		> 600			
Thickness under 2 kPa	mm	ISO 9863-1:05	7.5	4.4						

[1] Kaytech Engineered Fabrics (2014b)

[2] Kaytech Engineered Fabrics (2014c)

[3] Kaytech Engineered Fabrics (2010b)

[4] Kaytech Engineered Fabrics (2012)

5.2.4 Geomembranes

Two textured polyethylene geomembranes were investigated; one HDPE and one LLDPE. Both geomembranes were co-extruded, double-sided textured geomembranes with an asperity height of 0.4mm on each side and a nominal thickness of 1.5mm (GSE Environmental, 2015). These geomembranes were essentially impermeable geosynthetics that create a liquid barrier between contaminated fluids and surrounding soil and ground water. In landfill applications, without a geomembrane liner leachate would flow until it encountered groundwater posing the threat of pollution.

As illustrated in Figure 5-4, there was no visual difference between the two geomembranes tested. There are similarities and differences in their properties as seen in Table 5-3. The similarities allow for several variables to be eliminated during the investigation process. The two textured geomembranes were chosen such that the effects of geomembrane composition could be studied while maintaining the remaining geomembrane properties constant.

The most widely used geomembrane in the South African waste management industry is HDPE because this offers excellent performance for landfill liners and cover systems as illustrated in Figure 2-14. If greater flexibility than HDPE is required, then LLDPE is used because it has lower molecular weight resin that allows LLDPE to conform to non-uniform surfaces making it suitable for landfill caps (Shukla & Yin, 2006).

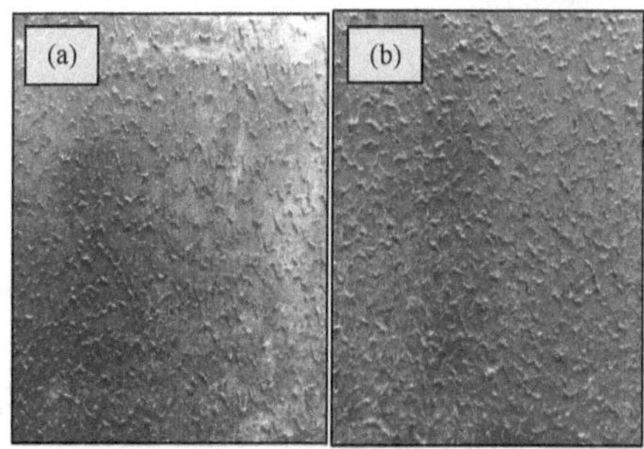

Figure 5-4: The two different types of geomembranes used (a) LLDPE geomembrane and (b) HDPE geomembrane

Table 5-3: Summary of HDPE and LLDPE geomembrane properties (GSE Environmental, 2015)

	Units	Standard	HDPE	LLDPE
Density	g/cm	ASTM D792	0.94	≤ 0.939
Tensile strength	N/mm	ASTM D638	16	18
Elongation	%	ASTM D638	100	250
Puncture resistance	N	ASTM D4833	400	300
Tear resistance	N	ASTM D1004	187	150
Thickness (nominal)	mm	ASTM D5994	1.5	1.5
Asperity (each side)	mm	ASTM D7466	0.4	0.4

5.3 Test apparatus and equipment

The equipment used for all tests in this research was the automated *ShearTrac-III* large direct shear box system (305 x 305 mm) designed and built by Geocomp Corporation Company, USA. It was capable of determining interface parameters between geosynthetic-geosynthetic, geosynthetic-soil and soil-soil. This apparatus was mainly selected because the ASTM D5321 recommended a large test box when shearing geogrids and many geocomposites. The large shear box was able to shear relatively large specimens with minimal edge effects. In addition, it has a shear displacement that is theoretically uniform across the width of each specimen tested.

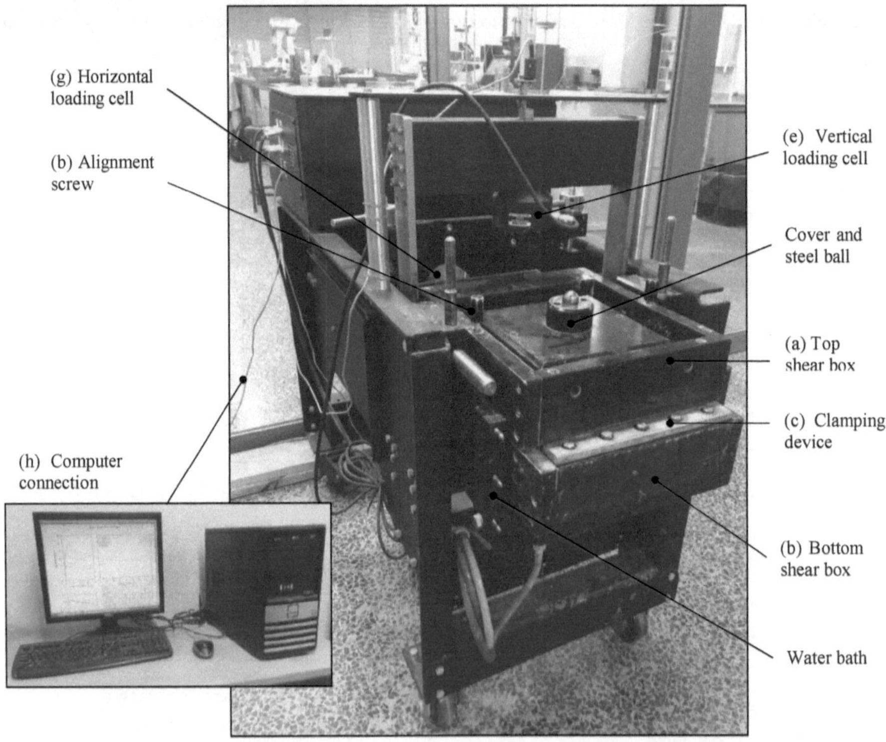

Figure 5-5: ShearTrac-III large direct shear apparatus and components

The shear box used consisted of the following components illustrated in Figure 5-5 and Figure 5-6:

a) Top box of 305 x 305 x 100 mm which was held in a static position during shearing,

b) Bottom box of 305 x 460 x 100 mm which connected to the top box using alignment screws. This split box could be moved relative to the top box and allowed a constant contact area between the geosynthetics being sheared for a maximum horizontal displacement of 70 mm. Both boxes were rigid enough to not distort during shearing of the specimen,

c) Clamping devices were located at the ends of both boxes. These devices allowed geosynthetics to be fastened to either the top or bottom box and did not interfere with the shearing surfaces within the shear box,

d) Metal substrate with measured dimensions of 305 x 460 x 100 mm was used when a soil sample was not required for the shear test,

e) Vertical loading cell applied a normal stress on the top box cover that rested on the geosynthetics after the two boxes were set in place. The vertical loading cell had a maximum load capacity of 450kPa,

f) Textured gripping plates designed to provide high friction that secured the test specimen in place were used. GCL interface shear testing required the gripping plates to minimise slippage and sliding while allowing the flow of water into and out of the test specimen,

g) Horizontal load cell applied shear stress to allow constant horizontal displacement to the bottom box and,

h) A computer was used to collect, store and display the data.

5.4 Research procedures

This study used two testing standards that measured total resistance to shear. The American Standard Testing Method (ASTM) D5321 that investigated geosynthetic/geosynthetic interfaces and the ASTM D6243 standard, which looked at geosynthetic clay liner (GCL) interfaces used.

These test methods attempted to model field anticipated conditions and performance of selected materials. Therefore the results obtained were limited to the application of projects with specific conditions considered during testing. This experimental work simulated the field conditions shown inTable 5-4: Table 5-4.

Table 5-4: Specific conditions considered during testing

Tested interfaces		Test rate (mm/min)	Pressure range (kPa)	Saturated
Material 1 (HDPE or LLDPE)	Material 2			
Geomembrane	GCL	0.5	25-300	Yes (24 hours)
Geomembrane	Geotextiles	1	25-300	No
Geomembrane	Geogrid	1	25-300	No

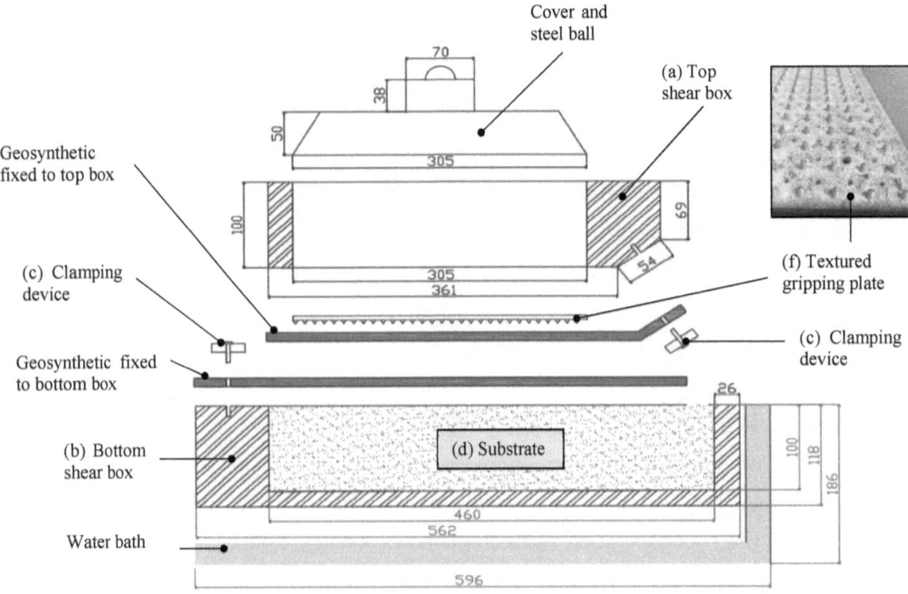

Figure 5-6: Exploded view of cross sectioned ShearTrac-III large direct shear apparatus and components (all dimensions in mm)

5.4.1 Test sample preparation

Geosynthetic sample preparation was done in accordance with the guidelines specified in ASTM D5321. The geosynthetics had to be cut into required sizes prior to testing. The size into which the geosynthetic samples were cut depended on which box the sample would be fixed onto. Samples that had to be fixed on the upper box were cut to 305 x 325 mm and those on the lower box were cut to 305 x 520 mm dimensions as demonstrated in Figure 5-7(a). Most of the geosynthetics were cut using a pair of scissors except the geomembranes that required a mechanical saw machine (Figure 5-7(b)) to be used.

Once samples were cut into the required sizes, 10 mm diameter gripping holes (five holes) were punched into one side of the geosynthetic sample using a mechanical hole puncher and hammer illustrated in Figure 5-7(c) and Figure 5-7(d) respectively. The holes allowed the sample to be fixed into place using the clamping devices.

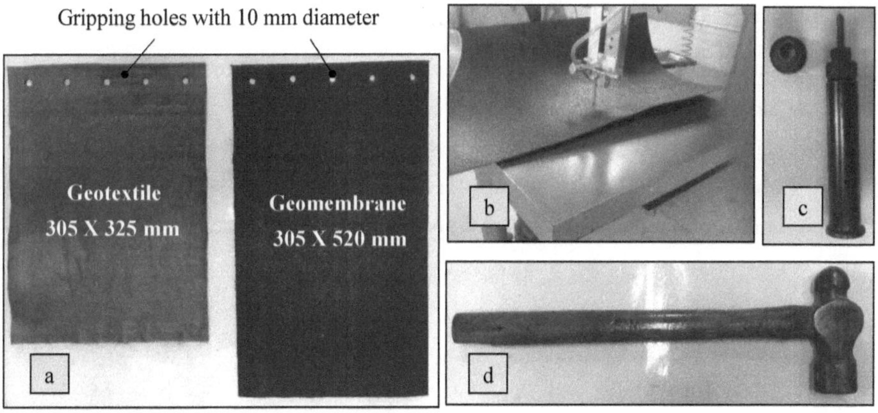

Figure 5-7: (a) Geosynthetic samples cut to box required dimensions, (b) Mechanical saw, (c) Mechanical hole puncher and (d) Hammer.

Some test samples required further preparation such as the GCLs. GCL samples had to be fixed to the top box, placed in position, then hydrated and consolidated to match field anticipated hydration (when it rains) and loading conditions (from waste). It is important to follow the same normal stress sequence for hydration in the laboratory as expected in the field because this sequence can affect the measured shear strengths (Fox & Stark, 2004). Thus, the specimens were hydrated for 24 hours under a confining pressure before shearing took place as specified in the ASTM D6243 standard. In this research, water was poured into the *ShearTrac-III* direct shear box water bath once the geomembrane/ GCL interface set-up was assembled.

The GCL was hydrated under a 20 kPa normal stress. This normal stress represented the possible initial waste load that would be imposed on the lining system. According to Fox & Stark (2004) hydration at low normal stress results in water being absorbed into the GCL. After hydration, the GCL clamping device was further tightened as it had loosened during hydration.

The water used to hydrate GCL combinations throughout this research was from four water sources located in the university Geotechnical Engineering Laboratory. Water has a molecular structure made of two parts hydrogen covalently bonded to one part oxygen.

Research has shown that the quality of water can affect the swelling capacity of GCLs (Thiel & Criley, 2005; CETCO, 2014). Table 5-5 illustrates pH values measured from the three laboratory water sources. The pH values indicate the water was slightly basic, lied within the recommended range specified by the Department of Water Affairs and Forestry (1996) (pH between 6.0 and 9.0) and contained no strong acids (pH < 2) or bases (pH > 12) which might damage the bentonite clay swell capacity (CETCO, 2014).

Table 5-5: pH of three laboratory water sources

Laboratory water sources	pH			
	Test 1	Test 2	Test 3	Average
Tap 1	8.46	8.47	8.45	8.46
Tap 2	8.19	8.16	8.19	8.18
Tap 3	8.48	8.52	8.50	8.50

5.4.2 Material and apparatus setup

In order to achieve apparatus setup, correct placement of prepared geosynthetic samples and additional large direct shear components (i.e. the metal substrate) was required.

As described in ASTM D5321, a metal substrate was used in the bottom box of the direct shear as a replacement for soil material. A test performed using a metal bottom fill material may influence the test results and may not stimulate field conditions as accurately as using a soil substrate. In this study, selection of a metal block was influenced by previous literature, which indicated that the use of this apparatus increased the accuracy and reproducibility of testing. Furthermore, test results that utilise a metal fill produce conservative direct shear resistance data when compared to soil substrate tests (Russell, et al., 1998; Parra, et al., 2012).

The geomembrane with dimensions 305 x 520 mm was placed flat over the substrate and fastened onto the bottom box with clamping devices as demonstrated in Figure 5-8(a). To ensure that the specimens tested did not pull out during the shearing process, the clamping devices were tightened with a torque ratchet (Figure 5-8(b)). A maximum torque of 38 Nm was applied on lubricated M10 stainless steel bolts. This torque value was determined using Table III-1 in the Appendix III, which ensured that the bolts were secured into place without damaging the female threads in the process.

Once the geomembrane was secure on the bottom box, the geosynthetic to be sheared against the geomembrane was fixed on the top box. The placement procedure of the geosynthetic depended on the type of material as elaborated below.

5.4.2.1   Geomembrane/ geosynthetic

The geosynthetic (geotextile or geogrid) cut to the dimensions of 305 x 325 mm was fixed onto the top box using the torque ratchet as demonstrated in Figure 5-8(c). The geotextiles and geogrid tested required no additional setup when fastened on top the top box.

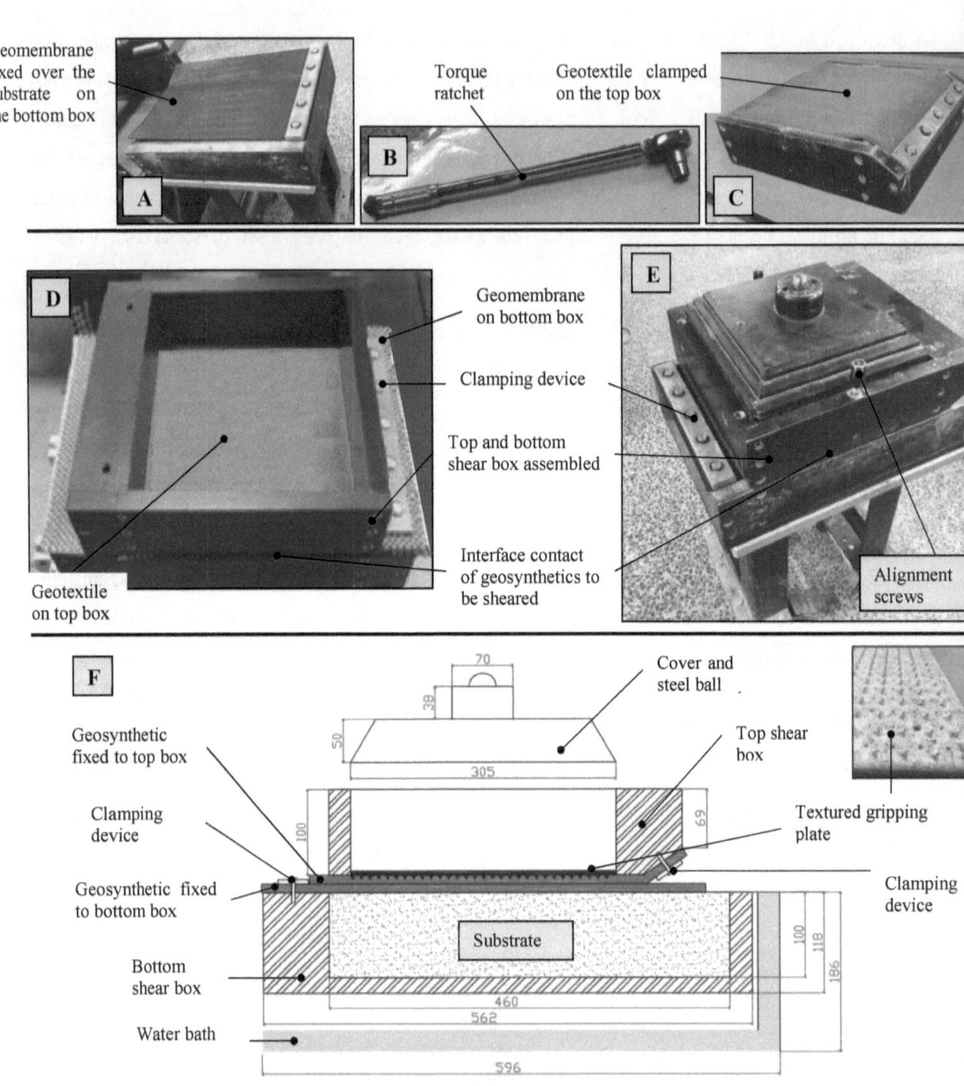

Figure 5-8: Assembling the direct shear box apparatus (all dimensions were in mm)

5.4.2.2   Geomembrane/ GCL

GCLs with dimensions of 305 x 325 mm were fixed on the top box similar to the geomembrane/ geosynthetic interface. Unlike geotextiles and geogrids, GCLs required an additional textured gripping plate to be used during testing as described in ASTM D6243. This gripping plate was located above the GCL (before the loading cover was placed). The plate held the specimen in position while preventing slippage between the shear box loading cover and GCL surface during the testing process. The steel gripping surface was made of 6.39 mm long nail spikes at 12.72 mm spacing mounted on a rigid metal plate as illustrated in Figure 5-9: Gripping plate details (all dimensions in mm). The gripping plate developed sufficient shear resistance to avoid interface failure between the geosynthetic material and the smooth under side of the loading cover. This forced shear to occur at the desired interface, prevented inaccurate peak values and large displacement shear strengths.

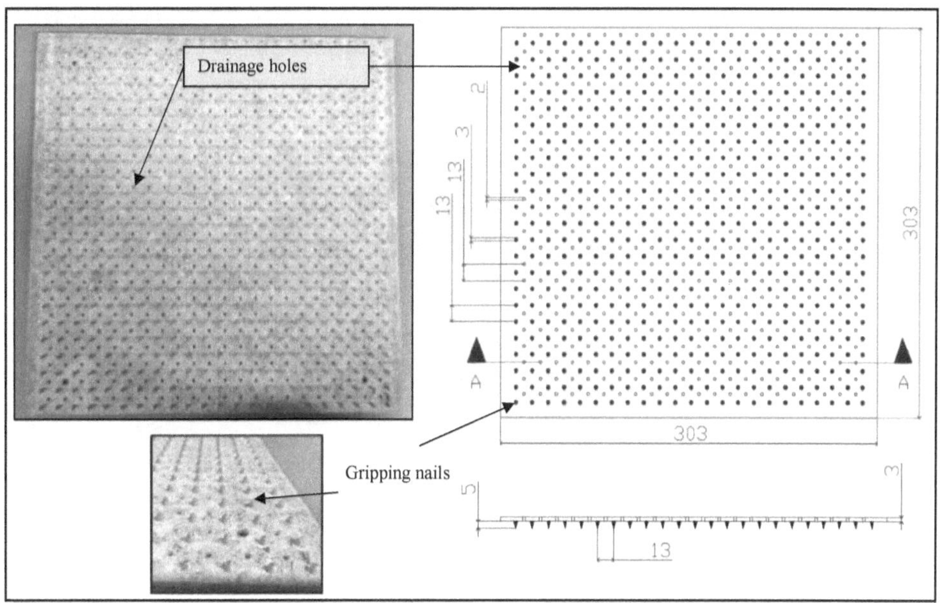

Figure 5-9: Gripping plate details (all dimensions in mm)

Once the geosynthetics were fixed firmly onto their respective boxes, the geosynthetic interfaces of interest were checked to make sure they were free of folds and any foreign material. When checks had been confirmed, the upper and lower boxes were assembled allowing the proposed geosynthetic surfaces to be sheared, to be positioned over each other (Figure 5-8(d)). Alignment screws were used to assist with positioning the two boxes into place. The loading cover along with a loading steel ball was then placed on top of the upper geosynthetic (or gripping plate in the case when GCLs are used) as illustrated in Figure 5-8(e) and Figure 5-8(f).

After the apparatus set-up was completed, the top and bottom boxes were moved to the starting position and pushed into the *ShearTrac-III* base container. The position of the boxes was then adjusted within the base container by lowering the vertical loading cell and moving the combined shear boxes horizontally, until the vertical loading device coincided with the steel ball placed on the loading cover as illustrated in Figure 5-10. When the set-up was in position, the normal stress was applied. Finally, nuts and bolts were used to secure the upper and lower shear boxes in the *ShearTrac-III* base container.

Figure 5-10: Vertical loading cell coinciding with the steel ball on the loading cover of the shear box

5.4.3 Testing procedure

Following the equipment set up, the *ShearTrac-III* software was opened and prepared to begin testing. The following test parameters were physically checked and entered on the software by the user for every test ran: (a) calibration values, (b) shear rate, (c) normal stress and (d) horizontal displacement details.

a) The shear device calibration was performed according to the manufacturer's manual. Geocomp Corporation Company recommended that the horizontal and vertical transducers used in the system should be calibrated for each test ran. The actual condition being sensed by the transducer needed to be physically entered by the tester during the calibration process. This allowed the program to calculate a calibration factor and offset value which was indicated by the sensor's output at zero condition (Geocomp corporation, 2012). Perfect zero readings were difficult to achieve due to the sensitive nature of the sensors.

b) The shear rates used were predetermined from previous literature. Many authors recommended shearing geosynthetic interfaces at a shearing rate of 1 mm/min and GCL interfaces at a lower rate of 0.1 mm/min to minimise bentonite extrusion (Triplett & Fox, 2001; McCartney & Zornberg, 2003; Fox & Kim, 2008). These shear rates were subsequently chosen and entered on the software (illustrated in Figure 5-11) to be used for the respective geosynthetics combinations in this research.

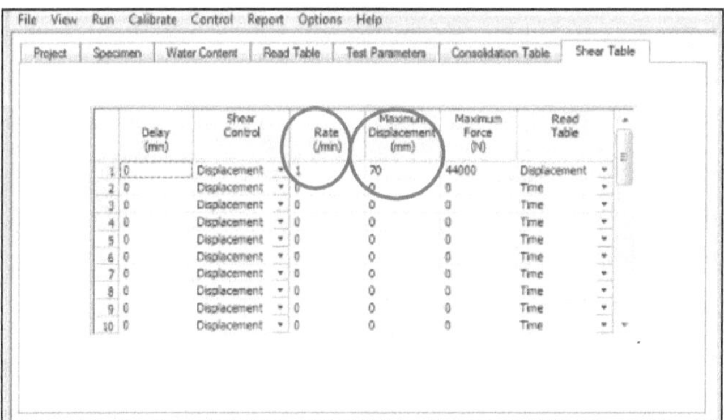

Figure 5-11: Screen shot of shear rate and horizontal displacement input window

c) The constant horizontal shear rate was used while a constant normal load was applied. The user as seen in Figure 5-12 entered the desired confining pressure on the software. The tests were run at six different normal loads namely 25, 50, 100, 150, 200 and 300 kPa. The large range of normal stresses was chosen to represent the varying loading conditions experienced by the liner system throughout the design life of the landfill. A relatively low normal stress immediately post-construction is expected, followed by an intermediate loading as waste placement proceeds and a long-term loading under full height of waste.

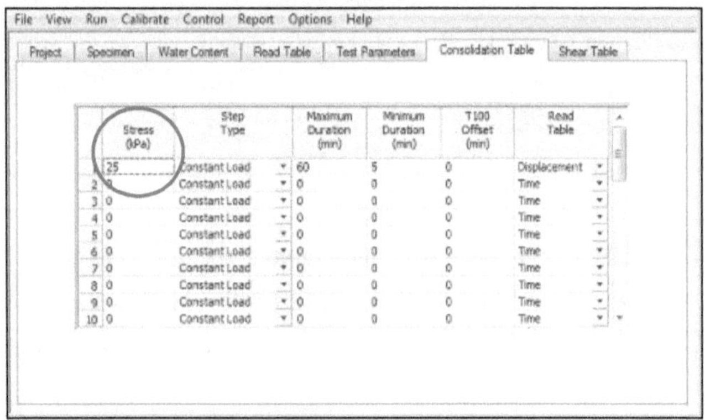

Figure 5-12: Screen shot of normal stress input window

d) The geosynthetics were sheared for a horizontal displacement of 70 mm at a constant rate of displacement, which was more than the ASTM D5321 and ASTM D6243 recommended minimum of 50 mm. This was to ensure that adequate shear box deformation was achieved to allow residual shear strength results to be recorded.

After entering the required data, the alignment screws were removed and the shearing phase was initiated. The *ShearTrac-III* device fixed the top shearing box in place and allowed the bottom box to move relative to the top box. A shear force was mobilized until sliding occurred between the geosynthetics. The length of the bottom box enabled the area of the failure surface not to decrease during shearing thus area corrections were not required when using the *ShearTrac-III*. This test was repeated at different normal stresses to enable the computer software to plot shear stress curves (at each normal load) and the limiting values of shear stresses against the applied normal compressive stresses used for testing.

5.4.4 Testing program

A total of 60 tests made up of 10 direct shear combinations were run. These tests were tested at varying normal loads namely 25, 50, 100, 150, 200 and 300 kPa. Thirty of the tests involved shearing against the HDPE textured geomembranes while the remaining thirty were sheared against the LLDPE geomembranes. The testing regime illustrating the details of the interface studies is presented in Table 5-6. Interpretation of the geosynthetic graphic symbols used is clarified in Table 5-7.

Table 5-6: Testing schedule for the large direct shear box

	Specimen	Test symbol	Test number according to applied normal stress for respective interfaces						Interface combination symbols
GCL	GCLA NW down (XA) / HDPE geomembrane	HDPE/NWD/XA	25	50	100	150	200	300	
	GCLA NW down (XA) / LLDPE geomembrane	LLDPE/NWD/XA	25	50	100	150	200	300	
	GCLB NW down (XB) / HDPE geomembrane	HDPE/NWD/XB	25	50	100	150	200	300	
	GCLB NW down (XB) / LLDPE geomembrane	LLDPE/NWD/XB	25	50	100	150	200	300	
Geotextile	GTA / HDPE geomembrane	HDPE/SFA	25	50	100	150	200	300	
	GTA/ LLDPE geomembrane	LLDPE/SFA	25	50	100	150	200	300	
	GTB / HDPE geomembrane	HDPE/CFB	25	50	100	150	200	300	
	GTB / LLDPE geomembrane	LLDPE/CFB	25	50	100	150	200	300	
Geogrid	Geogrid / HDPE geomembrane	HDPE/MGRID	25	50	100	150	200	300	
	Geogrid / LLDPE geomembrane	LLDPE/MGRID	25	50	100	150	200	300	

Table 5-7: Interpretation of geosynthetic graphic symbols

Geosynthetics	Symbols
LLDPE geomembrane	
HDPE geomembrane	
GCLA	
GCLB	
GTA	
GTB	
Geogrid	

5.5 Data analysis

The *ShearTrac-III* was connected to a computer unit which displayed test results in real time (Geocomp corporation, 2012). This data was exported to another software, Microsoft Excel, for data analysis.

The data was analysed in two different forms, namely the shear stress recorded as a function of the horizontal displacement of the moving section of the shear box and the six normal stresses plotted against shear stress. A line of best fit was drawn to connect the plotted points which established the Mohr-Coulomb failure plane. As a result, the gradient of the line gave the friction angle and the y-intercept gave the adhesion of the interface between the two materials tested. The data generally indicated peak strength that was higher than residual strength. Chapter 6 presents and discusses the results obtained using the equations below.

5.5.1 Vertical applied stress

Normal stress (σ_n): The vertical applied load from the loading cell divided by the test specimen contact area during shearing. It is computed based on Equation 5-1;

$$\sigma_n = \frac{N}{A} \ (kN/m^2)$$

Equation 5-1

Where,

N   - applied normal load in kN and,

A   - constant test specimen contact area in m^2.

5.5.2 Resisting shear stress

Shear stress (τ) is given by the resisting forces developed along the interface plane being considered between the geosynthetics sheared. It is a ratio of the force divided by the interface contact area as given in Equation 4-1 as:

$$\tau = \frac{F}{A} \ (kN/m^2)$$

Where,

F   - shearing force generated by the horizontal loading cell which causes the lower split box to move relative to the upper box measured in kN and,

A   - constant test specimen contact area in m^2.

5.5.3 Mohr-Coulomb failure envelope

Using the peak and residual shear strength values and corresponding normal stresses allows for the Mohr-Coulomb failure envelopes to be developed. The straight-line response is a result of the following Equation 5-2:

$$\tau = C_a + \sigma_n tan\phi_A \ (kN/m^2)$$

Equation 5-2

Where

τ - shear stress resistance acting along the geomembrane/ geosynthetic surface in kN/m^2,

σ_n - normal stress on the shear plane in kN/m^2,

C_a - adhesion of geomembrane to opposing surface in kN/m^2 and,

ϕ_A - friction angle between the geomembrane and opposing surface in degrees (°).

The data analysis process allows interpretation of research methodology summarised in Table 5-8

5.6 Quality assurance

To ensure quality in the results produced, certain measures were strictly adhered to. Understanding and addressing numerous factors that could not be controlled increased statistical validity and comparability of results. These aspects were considered from the selection of the sample to testing procedures and analysis of results. These aspects are highlighted below:

- The geosynthetic samples for testing were properly selected and cut to eliminate those sections with damages or faults on them. This ensured that samples with a uniform surface roughness were used so that false shear strength results were not achieved.

- Both geomembranes, GTA, GTB and the geogrid were cut to allow shear in the weakest cross machine direction to avoid overestimating shear strength of the products in the field. The supplier indicated the machine direction on the material. For GCLA and GCLB, the weakest direction did not exist therefore specimens were cut depending on the dimensions of the original samples received.

- During hydration of a GCL, the time for the dewatering process and beginning to shear was kept at a minimum (Approximately 5 minutes). Immediately after all the water had been drained from the shear box bath, shearing of the GCL- geomembrane interface commenced thus not allowing time for high evaporation of water.

- The repeatability of the experimental procedure and results was verified by conducting 3 experiments under the same testing conditions and confirming that their results were repeatable.

- In addition to the above measures, a fresh geosynthetic specimen was used every time a new experiment was set up, all equipment was properly calibrated and all tests were conducted under similar physical and climatic (air-conditioned laboratory at 22 °C) conditions.

Table 5-8: Summary of research methodology

Step	Activity	Description
1.	**Sample preparation**	• Geosynthetic preparation: the sample was cut to a specific size depending on where it would be clamped, namely the top or bottom shear box. Geosynthetic test samples had clamping holes punched into them, which were used to fix the specimens into place. • GCLs had to be hydrated with water for 24 hours under a load of 20kPa prior to the shearing process. A water bath was created to allow the GCL to be submerged in water.
2.	**Assembly of apparatus**	• A metal substrate was placed into the bottom box and the geomembrane sample was fixed over the substrate, • A geosynthetic to be sheared with the geomembrane was fixed onto the top box, • The top shear box was placed over the bottom box with the assistance of alignment screws, • If a GCL was being sheared, a textured gripping plate was placed over the geosynthetic in the top box, • The loading cover and steel ball was place above the top geosynthetic, • The assembled mechanism was pushed into the shearing device and • The vertical loading cell was lowered until it coincided with the steel ball placed on the loading cover.
3.	**Computer input**	• *ShearTrac-III* software was used to facilitate the testing procedure, • Test details were entered (test number, date, tester details and sample details), • Calibration data values were entered, • Desired shearing rate was entered, • Desired normal stress load was entered and • Desired maximum horizontal displacement was entered.
4.	**Data analysis**	• The test was repeated for six different normal loads to produce a Mohr Coulomb's failure plane, • A shear stress against horizontal displacement graph and • A graph indicating normal stress against shear stress was plotted. A line of best fit between the values produced adhesion and coefficient friction angle design values.

5.6.1 Repeatability of results

It is desirable practice to check the reproducibility and reliability of results conducted under identical test conditions. In Figure 5-13 and Figure 5-14, the relationship between the shear stress and horizontal displacement of HDPE geomembrane/ GTB and LLDPE geomembrane/ GCLA interface under constant normal stresses of 25 kPa and 200 kPa respectively, are presented. It was assumed that the selected interfaces were reasonable representation of common geomembrane/ geotextile and geomembrane/ GCL interfaces. The three curves in each figure represent the shear behaviours of the interfaces during the first, second and third shear tests for identical test specimens. The figures confirm that the machinery used and the test procedure adopted produced consistency among results and displayed repeatability. The precision determined from the multiple test results showed that ultimate shear stresses for the three tests in each graph were nearly equal although slight variations were noted along the curves. Generally, it was noted that the peak shear stresses in the HDPE geomembrane/ GTB combination were achieved at relatively similar horizontal displacements of approximately 3 mm.

The analysis shown in Table 5-9 indicated that the maximum deviation from the peak and residual averages was 2.4 % and 1.27 % respectively and from Figure 5-9 was 1.49 %. This level of accuracy was acceptable since it falls under the maximum suitable standard deviation of 5 % from the mean stresses that was considered sufficient to affect the results (Cochran, 1963). Hence, it was confirmed from the consistency of the results obtained that the experimental procedure followed was reproducible, which suggested that all the tests conducted throughout the study were repeatable.

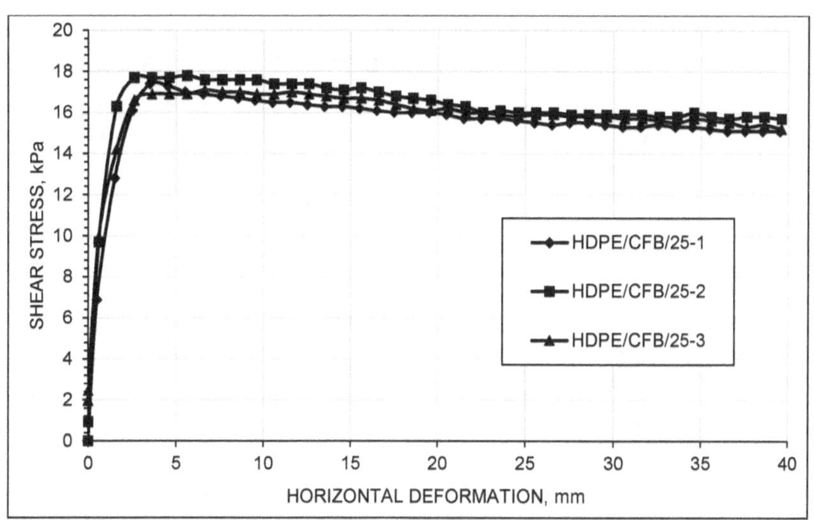

Figure 5-13: Graph of shear stress against horizontal displacement for HDPE geomembrane against GTB

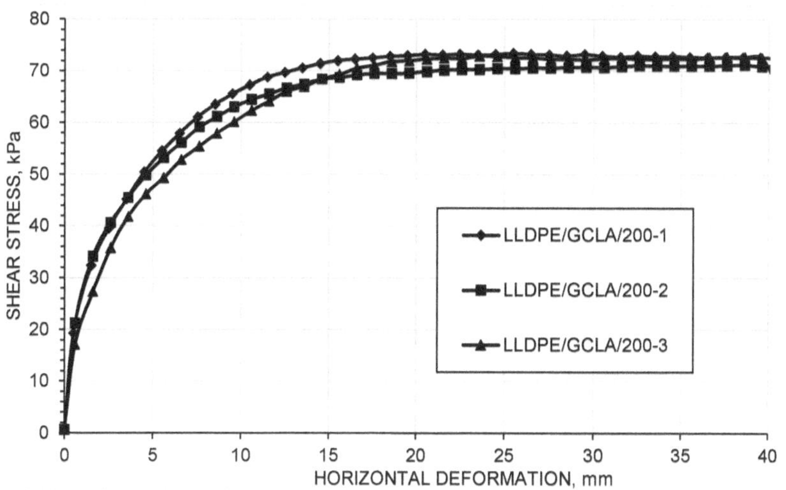

Figure 5-14: Graph of shear stress against horizontal displacement for LLDPE geomembrane against GCLA

Table 5-9: Repeatability results analysis for HDPE geomembrane/ GTB interface

Test specimen	Peak stress (kPa)	Average peak stress (kPa)	Deviation from average		Residual stress (kPa)	Average residual stress (kPa)	Deviation from average	
			kPa	%			kPa	%
HDPE/CFB/50-1	17.40		-0.07	0.40	15.53		-0.20	1.27
HDPE/CFB/50-2	17.90	17.47	0.43	2.40	15.90	15.73	0.17	1.08
HDPE/CFB/50-3	17.10		-0.37	2.16	15.77		0.04	0.25

Table 5-10: Repeatability results analysis for LLDPE geomembrane/ GCLA interface

Test specimen	Residual stress (kPa)	Average residual stress (kPa)	Deviation from average	
			kPa	%
LLDPE/GCLA/200-1	72.67		0.61	0.85
LLDPE/GCLA/200-2	70.98	72.06	-1.08	1.49
LLDPE/GCLA/200-3	72.53		0.47	0.65

6. Results and discussions

6.1 Introduction

Direct shear tests results conducted on High Density Polyethylene (HDPE) and Linear Low Density Polyethylene (LLDPE) geomembranes against two GCLs, two geotextiles and a geogrid were presented and discussed in this chapter. With the objective of determining the effect a geomembrane polymer type has on interface friction characteristics, the results were initially illustrated and discussed in the form of interface shear stress development versus horizontal displacement diagrams. This was followed by an in-depth analysis of the relationship between volume change and the time of shear. Lastly, a detailed examination of the maximum interface shear stress (as well as the residual interface shear stress) plotted against the applied normal pressure was given.

6.2 Shear stress - horizontal displacement relationship

6.2.1 Effect of geomembrane composition on shear stress

The shear stress versus horizontal displacement relationships of materials tested according to the direct shear test program (Table 5-6) were shown in Figure 6-1. For each geosynthetic material sheared against a geomembrane, a separate graph was produced i.e. Figure 6-1(a) to Figure 6-1(e). In each diagram, i.e. (a) to (e), there were two sets of responses representing HDPE geomembrane/ geosynthetic and LLDPE geomembrane/ geosynthetic interface tests as indicated in the legend at the bottom right of the figures. Each of the interface tests were conducted at six different normal pressures namely at 25 kPa, 50 kPa, 100 kPa, 150 kPa, 200 kPa and 300 kPa. Hence, there were six curves of the same line type in each graph (i.e. (a) to (e)). The large range of normal stresses was chosen to represent the varying loading conditions experienced by a typical landfill cover and liner system throughout the design life of a containment facility. A relatively low normal stress immediately post-construction could be expected followed by an intermediate loading as waste placement proceeded on the base liner or as the cover liner system becomes fully operational.

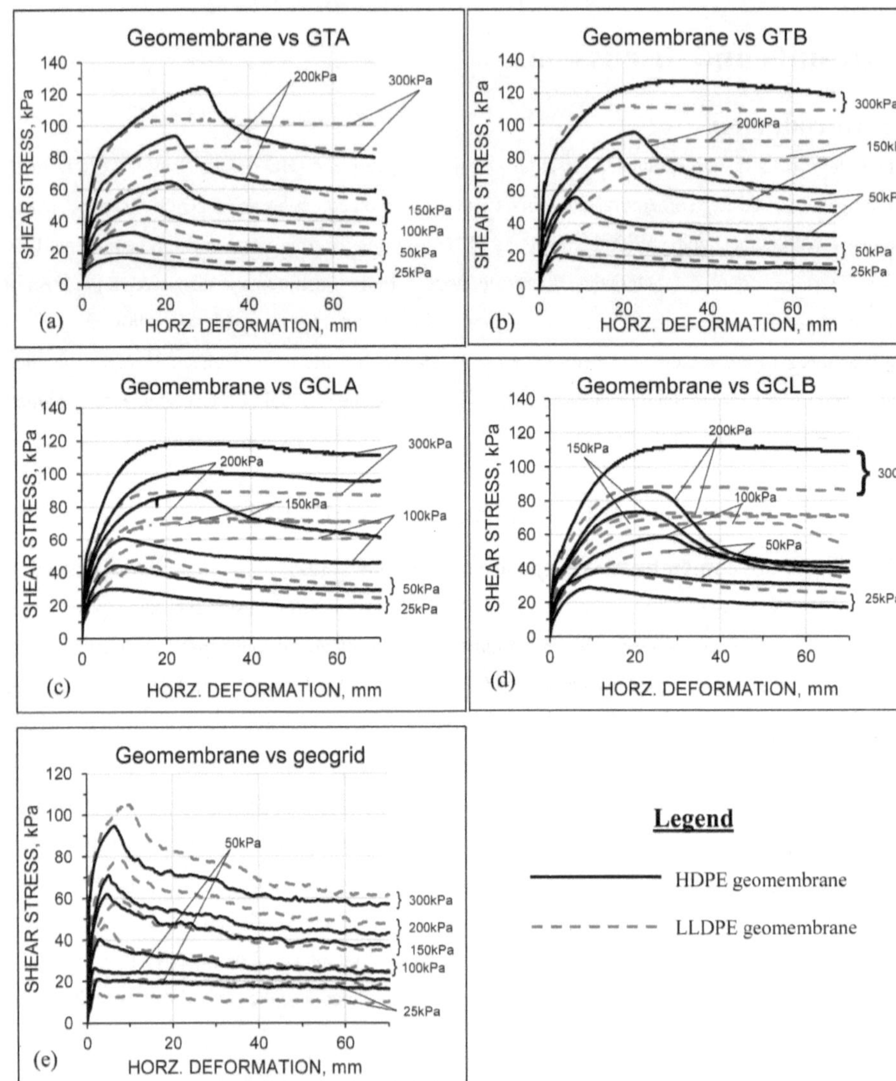

Figure 6-1: Shear stress versus horizontal displacement of geomembrane/ geosynthetic interfaces

114

Generally, it was observed from Figure 6-1 that the rate of development of shear stress with horizontal displacement increased as the normal pressure increased. Moreover, the shear stresses of all geomembrane/ geosynthetic interfaces increased rapidly with increasing horizontal displacement before reaching a peak shear strength. Once the peak strength was reached, a gradual reduction of the shear strength occurred with further increasing horizontal shear displacement, a phenomenon known as strain softening. In several cases, the shear stresses finally reached a steady-state value with increasing horizontal displacement. This value was considered the residual shear strength. In general, the LLDPE curves did not depict this pattern. The LLDPE strain softening curves were almost horizontal and showed no distinct residual strength. The same horizontal strain softening pattern was observed on several HDPE interfaces sheared at 300 kPa normal stress.

The shear stress – horizontal displacement relationship results obtained in this investigation were very similar to published test results reported in literature (Bhatia & Kasturi, 1995; Russell, et al., 1998; Blond & Elie, 2006; Fox & Kim, 2008; Bacas, et al., 2015) in respect to the shape of the stress-displacement curves, the peak shear strength mobilised and the residual shear strength mobilised at similar interfaces tested.

The shear strength patterns along HDPE and LLDPE geomembrane interfaces were non-identical and depended primarily on the properties of both the upper and lower materials. Thus, the shear strength versus horizontal displacement curves had been utilized to describe this phenomenon and determine the relationship at pre-peak, peak, strain softening and residual stages of the different geomembrane interfaces.

6.2.1.1 Pre-peak

The initial rapid increase of the shear stress and horizontal displacement relationship of HDPE and LLDPE geomembrane interfaces seen in Figure 6-1 could be classified as the pre-peak stage and represented using a mathematical model (Clough and Duncan, 1971). Generally, the shape of the curve was non-linear before reaching the peak shear strength thus the model often provided an accurate approximation of the interface behaviour at constant normal stress by illustrating the initial shear modulus (Ksi) and tangent shear modulus (Kst) of the interface as demonstrated in Section 4.4.1.

The practical significance of these parameters highlights the stiffness of the tested materials at different stages of shear. The extent to which the tested materials resist deformation when they experience initial shear stress is represented by Ksi, within the elastic region. The Kst is useful in describing the behaviour of materials that have been stressed beyond the elastic region. When plastic deformation occurs there is no longer a linear relationship between stress and strain as there is for elastic deformation. The Kst quantifies the initial softening of the tested materials that generally occurs when it begins to yield.

For illustration purposes, data from an interface shear test of HDPE geomembrane/ GTA test was used to display how hyperbolic parameters were determined and how the basic aspects of the mathematical model were plotted. Figure 6-2 (determine a and b parameters), Figure 6-3 (represent Kst modulus) and Figure 6-4 (represent mathematical approximation) were used in this presentation. Similar graphs were demonstrated in Appendix IV for all geosynthetic combinations tested which enabled Ksi and Kst values to be calculated as shown in Table 6-1.

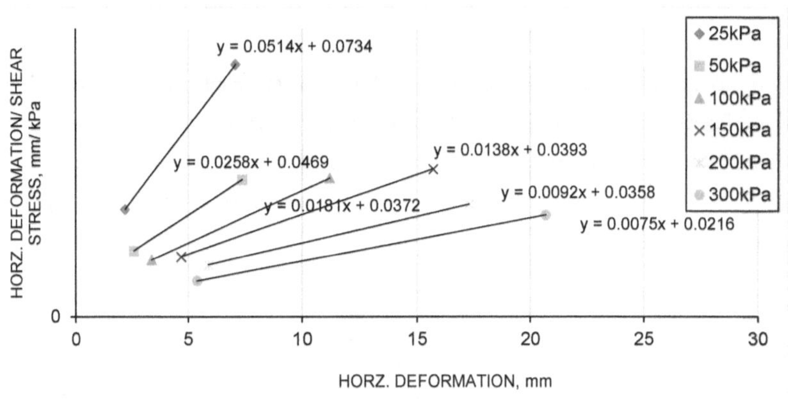

Figure 6-2: Determination of hyperbolic parameters for a HDPE geomembrane/ GTA interface

Based on Table 6-1, the Ksi modulus appeared to generally increase with the applied pressure for each geomembrane type sheared against the different geosynthetics (i.e. GTA, GTB, GCLA, GCLB and the geogrid). High Ksi values indicated a higher state of inflexibility of the sheared interface. This behaviour could be attributed to the increased confining pressure influencing the material rigidity of the polyethylene geomembranes and geosynthetics tested. Greater interaction caused by the high normal stresses prompted an increase in the inability for the surfaces to be sheared thus resulting in a measure of increased material stiffness.

There seemed to be an evident trend demonstrating that at higher normal stresses (greater than 200 kPa) the Ksi values for LLDPE geomembrane interfaces become close or higher to the Ksi values of HDPE geomembrane interfaces. This was true for geotextile and GCL interactions. This behaviour could be influenced by the large applied confining pressures that caused the flexible LLDPE geomembrane asperities to tightly interact with the adjacent geosynthetic thereby developing increasing resisting interface shear stresses during initial shear.

Generally, the Ksi values for geogrids were significantly higher possibly due to high deformation resistance between the geomembrane-geogrid combination. This may imply that during initial shear, the geogrid 'gripped' the geomembrane asperities thus increasing shear resistance.

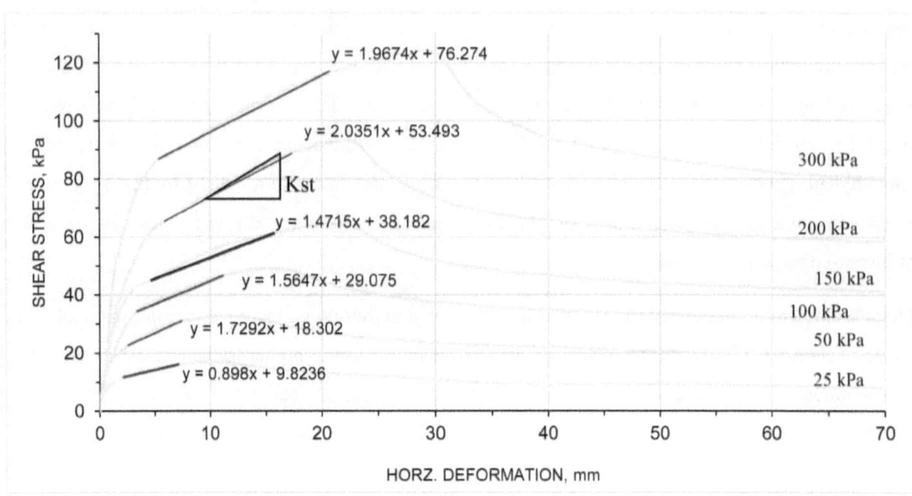

Figure 6-3: Determination of tangent shear modulus (Kst) for HDPE geomembrane/ GTA interface

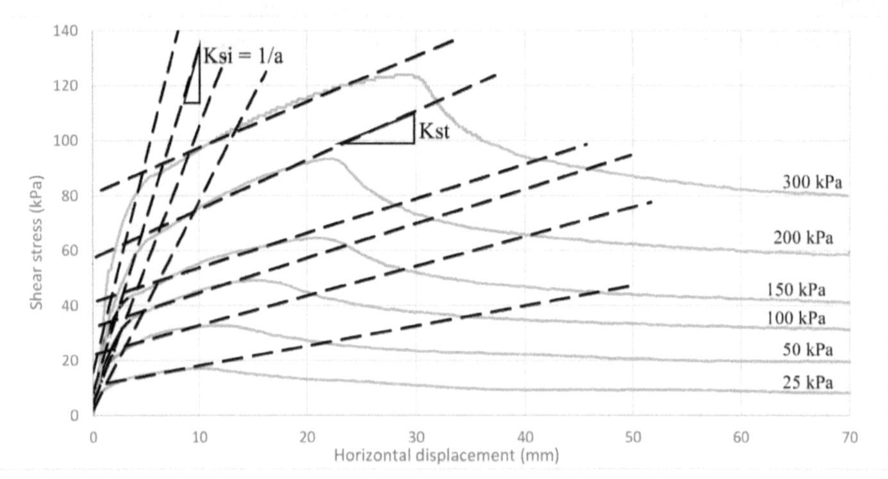

Figure 6-4: Mathematical model approximation of plotted direct shear test data of HDPE geomembrane/ GTA

HDPE geomembrane interfaces showed large Kst values when compared to Kst values of LLDPE geomembranes tested against majority of geosynthetics. This was true for all HDPE geomembrane/ geosynthetic interfaces except when sheared with GTA and GTB interfaces at confining pressures greater than 200 kPa. The high frictional resistance gradient or Kst of the HDPE geomembrane/ geosynthetic interfaces could be attributed to the stiffness of the geomembrane. When the HDPE geomembrane interlocks to the adjacent geosynthetic, the HDPE asperities are able to maintain their original shape during early shear. This leads to improvement in the shear strength and mechanical behaviour during the tangent shear modulus approximation. On the other hand, flexible LLDPE asperities are assumed to plastically deform during initial shear resulting in reduced tangent shear moduli. It was anticipated the needle punched fibres of GTA and GTB allow the LLDPE geomembrane roughened surface to interlock tightly at high normal stresses (greater than 200 kPa) resulting in limited asperity deformation and increased interface stiffness during initial shear.

Table 6-1: Initial shear modulus (Ksi) and tangent shear modulus (Kst) of different HDPE and LLDPE geomembrane combinations tested

	KN /m³	HDPE geomembrane		LLDPE geomembrane			HDPE geomembrane		LLDPE geomembrane			HDPE geomembrane		LLDPE geomembrane	
		Ksi	Kst	Ksi	Kst		Ksi	Kst	Ksi	Kst		Ksi	Kst	Ksi	Kst
GTA — Normal stress (kPa)	25	13.62	41.92	10.13	62.35	GCLA	27.03	70.91	11.10	60.6	Geogrid	13.19	83.21	56.50	80.08
	50	21.32	59.96	11.25	58.65		24.27	72.50	18.52	60.71		56.82	84.45	75.76	80.19
	100	26.88	57.42	10.32	55.97		31.06	74.30	21.88	55.45		75.19	82.34	86.21	80.14
	150	25.45	55.80	16.84	55.42		27.47	65.13	27.40	62.43		86.96	82.02	69.44	80.06
	200	27.93	63.83	22.68	64.60		26.81	69.22	25.58	67.14		74.07	82.57	112.4	78.74
	300	46.30	63.06	45.25	72.95		36.76	74.86	36.36	72.74		200	82.02	192.3	78.94
GTB — Normal stress (kPa)	25	12.95	73.00	8.56	58.74	GCLB	11.99	64.06	13.99	59.21					
	50	13.50	73.96	7.77	63.44		31.75	55.10	9.60	45.93					
	100	22.73	75.48	11.81	52.98		13.19	49.93	13.81	50.64					
	150	14.10	69.82	23.87	60.29		16.03	67.98	14.43	59.88					
	200	23.81	66.17	29.94	68.12		15.75	67.65	19.88	57.06					
	300	47.62	67.42	69.44	78.10		22.47	69.35	22.42	68.70					

It was observed from Figure 6-1, Figure 6-4 and Table 6-1 that the initial shear stiffness for each interface tested varied depending on the magnitude of confining stress employed. Furthermore, the findings demonstrated that the tangent shear modulus mainly varied with the horizontal displacement.

From the values in Table 6-1, logarithmic graphs of Ksi/γ_w and σ_n/Pa (attached in Appendix V) were used to calculate the material properties for the different geomembrane/ geosynthetic combinations. Table 6-2 illustrated shear coefficient (K) and modulus exponent (n) properties obtained for HDPE and LLDPE geomembranes sheared against various geosynthetics. The practical significance of the K value highlights the extent at which the tested materials resist deformation. The value of n describes the material behaviour influenced by normal stress loading. These K and n material values were comparable to those produced in literature shown in Table 6-3.

Majority of HDPE geomembrane interfaces were considered to be more rigid than LLDPE surfaces due to the slightly larger K values obtained. A geomembrane interface with high rigidity would be able to provide greater resistance at initial shear, when sheared against geotextiles and GCLs. Moreover, reasonable flexibility at the pre-peak region was observed for LLDPE geomembranes interfaces which was represented by a distinct pattern of higher n parameters. A higher value of n indicates dependency of tangent modulus on normal stress in the pre-peak region. Limited ability of ductility was anticipated for tested interfaces that obtained n values close to zero, mainly seen at HDPE geomembrane/GCL combinations.

The above was true for geomembrane/ GCL and geomembrane/ geotextile interfaces tested; however geomembrane/ geogrid combinations indicated the opposite. With higher rigidity for LLDPE geomembranes and greater flexibility for HDPE geomembranes.

Table 6-2: Hyperbolic interface model parameters shear coefficient (K) and modulus exponent (n) at pre-peak region

Geosynthetic	HDPE		LLDPE	
	K	n	K	n
GTA	2.563	0.40	1.678	0.54
GTB	2.030	0.43	1.867	0.84
GCLA	2.917	0.096	2.240	0.43
GCLB	1.781	0.07	1.538	0.24
Geogrid	6.545	0.88	9.213	0.39

Table 6-3: Literature hyperbolic interface model parameters at pre-peak region

Interface	K	n	Reference
HDPE GMX-GT	5.314	0.18	(Jones and Dixon, cited in Seo, et al., 2003)
HDPE GMX-GCL	1.540	0.65	(Triplett and Fox, 2001)

6.2.1.2 Peak

Following the pre-peak stage, the shear stress versus displacement graph reached the maximum stress also known as the peak shear stress. At this stage, the influence of the geomembrane composition on interface peak strength was investigated by comparing the magnitudes of peak shear stresses of HDPE geomembrane and LLDPE geomembrane sheared against GTA, GTB, GCLA, GCLB and a geogrid as illustrated in Table 6-4. Table cells lightly shaded in grey show the geomembrane with the highest peak value when sheared against the same geosynthetic material.

From Table 6-4, it was evident that at normal confining pressures lower than 100 kPa, the LLDPE geomembrane/ geosynthetic interfaces presented larger peak values than HDPE geomembranes sheared against majority of the geosynthetics. This behaviour could be a result of the interaction of the contact surfaces mainly between the superficial filaments of the geotextile or GCL planes and the asperities of the geomembranes. The flexible LLDPE geomembrane surfaces would be able to adjust and fit closely to the adjacent geosynthetic shape, which enabled the asperities to embed and interlock tightly within the needle-punched filaments thus increasing the shear stress achieved, an attribute beneficial for geomembranes on landfill capping systems. Smaller HDPE interface shear strength at low normal stresses could be contributed by rigid geomembranes surfaces and asperities not developing high interlocking mechanism on a superficial level as much as the LLDPE geomembranes.

Table 6-4: Effect of geomembrane composition on peak shear stresses

	Normal stresses	HDPE geomembrane	LLDPE geomembrane		HDPE geomembrane	LLDPE geomembrane		HDPE geomembrane	LLDPE geomembrane
GTA	25	17.1	25.2	GCLA	29.9	44.3	Geogrid	21.1	14.8
GTA	50	32.8	41.6	GCLA	44.1	49	Geogrid	26.4	21.5
GTA	100	49.2	63.1	GCLA	60.5	60.7	Geogrid	40.4	47
GTA	150	64.5	76	GCLA	87.9	71.4	Geogrid	62	58.5
GTA	200	93.4	87.1	GCLA	101	73.4	Geogrid	71.1	78.6
GTA	300	124	104	GCLA	118	89.1	Geogrid	94.6	106
GTB	25	20.3	21.7	GCLB	36.7	47.9			
GTB	50	31.7	40.6	GCLB	38.7	50.3			
GTB	100	56	73.4	GCLB	64.3	66.9			
GTB	150	83.4	79.1	GCLB	73.4	71.3			
GTB	200	95.6	90.7	GCLB	85.6	73			
GTB	300	127	112	GCLB	112	88.5			

For normal stresses higher than 100 kPa, comparatively lower peak shear values for flexible LLDPE geomembrane interfaces were obtained. This behaviour could be influenced by the large applied confining pressures that compress the LLDPE asperities resulting in a decreased ability to interlock between geosynthetic interfaces. High peak stress values of HDPE geomembrane interfaces could be an effect of continued increase in tight embedment of undamaged geomembrane asperities and filament surfaces as confining pressures increase, thus developing a continued rise in resisting interface shear stresses, similar to the trend observed by McCartney, et al. (2002).

Unlike geotextile and GCL surfaces sheared against HDPE and LLDPE geomembranes, geogrid interfaces demonstrated diverse peak stress values. It was more difficult to distinguish a clear trend between the geogrid and geomembrane interactions possibly due to the local stiffness of the interface increasing solely through friction of the textured geomembranes. Geogrids are commonly designed with a friction and interlock function, and limited interface friction parameters are achieved when both geogrid design functions are not implemented. Generally, higher peak values were observed for HDPE geomembranes at low confining pressure (less than 50 kPa) and LLDPE geomembranes at high normal stresses (more than 200 kPa).

From respective shear stress – horizontal displacement geomembrane/ geosynthetic responses (refer to Figure 6-1), the effect of geomembrane composition on horizontal displacement behaviour of various geomembrane /geosynthetic interfaces is illustrated in Figure 6-5. As shown in Figure 6-1, the maximum shear stress occurred after a certain amount of horizontal displacement had taken place and the displacement at peak shear stress varied depending on the geosynthetic materials tested. The data plots in Figure 6-5 show the relationship of the horizontal displacements required to fully mobilise the maximum shear stress for different geomembrane/ geosynthetic interfaces under specified confining pressures.

Based on Figure 6-5, for any particular HDPE and LLDPE geomembrane combination, it can be seen that the trend formed by the horizontal displacement required to fully mobilise the peak shear stress differed subject to the normal pressure applied. It can be observed, HDPE geomembrane interfaces against GTA, GTB and the geogrid indicated an increase in horizontal displacement as the confining pressure increased. Similarly, the LLDPE geomembrane interfaces against the geogrid demonstrated an enlargement in horizontal displacement as the confining pressure increased. It is possible that an increase in normal pressure exerted on a geomembrane/ geosynthetic interface results in an improvement of the stiffness and degree of interlock between the contact surfaces thereby developing a rapid increase in resisting interface shear stresses at the pre-peak stage. For instance, as the layers of waste that occupy a landfill base liner increase, it is anticipated that the resistance between the above mentioned interfaces would increase. Geotextile and GCL interfaces sheared against LLDPE geomembranes demonstrated that after a maximum horizontal displacement was reached (between 100 kPa and 200 kPa), there was a decrease in horizontal displacement as the normal stress increased. This behaviour could be attributed to the normal stress limiting the plastic deformation of the flexible LLDPE geomembrane at high normal stresses. Therefore, significantly increasing the stiffness of the interaction mechanisms between the LLDPE geomembrane and the adjacent geosynthetic. In addition, HDPE geomembrane/ GCL combinations showed no specific trend between horizontal displacements mobilised at maximum peak stress and various confining pressures.

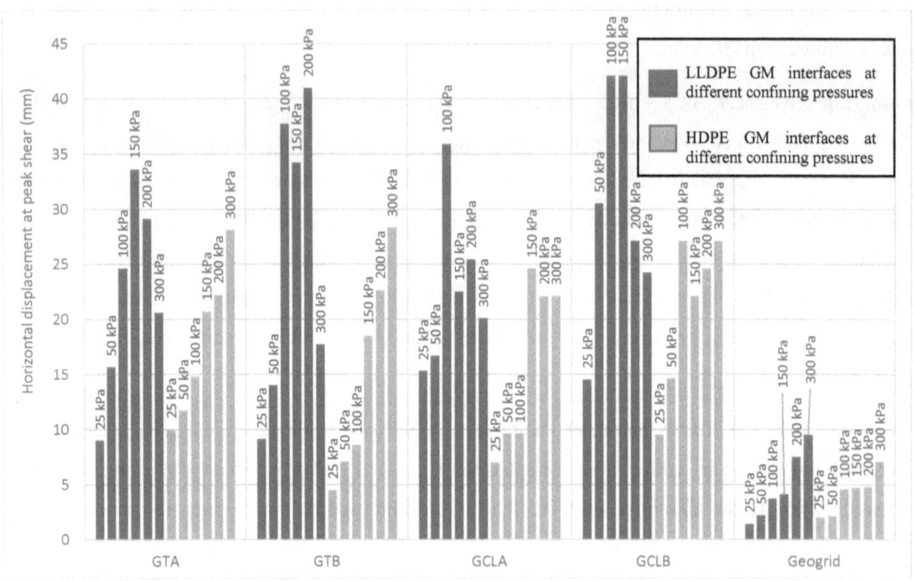

Figure 6-5: The effect of geomembrane composition on horizontal displacement

As a final comparison, it is noted that the horizontal displacement at any particular confining pressure required to achieve maximum shear at the geomembrane/ geogrid interface was well below those of other geomembrane/ geosynthetic interfaces for both HDPE and LLDPE geomembranes. This may imply that the rigid geogrid surface does not 'grip' the geomembrane asperities, thus reducing the interface mobilisation interaction to a minimum unlike geotextile and GCL interfaces.

6.2.1.3   Strain softening

In several cases, once the interface peak stress is reached, the test results show the shear strength decreasing with increasing horizontal displacement, a phenomenon known as strain softening. In the softening stage, the shear strength initially demonstrates a sharp reduction with a slow increase in the shear displacement and then shows a gradual strength drop. This interface degradation behaviour can be seen in shear stress – horizontal displacement curves for mainly HDPE geomembrane interfaces in Figure 6-1. In general, the LLDPE curves did not depict this pattern. The LLDPE strain softening curves were almost horizontal and showed no distinct residual strength. The same horizontal strain softening pattern was observed on several HDPE interfaces sheared at 300 kPa normal stress.

The primary factor affecting the shear strength of the interface remains the nature of the contact surfaces. Comparing the HDPE geomembrane/ geosynthetic with LLDPE geomembrane/ geosynthetic interface results in Figure 6-1, the strain softening region was significantly influenced by the confining pressures as highlighted by Stark et al. (1996). In the current study, the HDPE and LLDPE geomembrane interfaces had large variations in the reduction percentage of peak shear strength to residual strength for different normal stresses.

Large strength loss may be influenced by the interaction mechanisms between the asperities of the textured geomembrane and the adjacent geosynthetic during the shearing process. Interbedding and interlocking contact between interfaces cause the shear strength to reach its peak. It is possible that after the peak strength is mobilised, the geomembrane asperities pull out, tear, untangle, deform and orientate the adjacent geosynthetic filaments in the direction of shear which could result in the interbedding and interlocking mechanism degradation until a steady state is reached (Russell, et al., 1998; Kim, 2006). The interaction mechanism and decrease in interface shear strength mainly depends on the type of geosynthetic and geomembrane utilised.

The degradation in shear strength at each normal stress applied (25 to 300 kPa) was shown by utilising a dimensionless parameter referred to as the sensitivity ratio which is measured by dividing the peak strength (peak τ) by the residual strength (residual τ) as demonstrated in Equation 6-1. A sensitivity ratio close to one implied a non-significant strength loss.

$$Sensitivity = S_\tau = \frac{\tau_{peak}}{\tau_{residual}}$$

Equation 6-1

Figure 6-6 illustrates the variations in sensitivity throughout the range of normal stresses used in this study. In general, the sensitivity of both LLDPE and HDPE interfaces showed similar trends at low normal stresses. At higher normal stresses (in the range of 100 to 300 kPa), the confinement resulted in an increase in HDPE geomembrane sensitivity and a decrease in the LLDPE geomembrane sensitivity.

Similarly it was also observed in Figure 6-1, that a number of LLDPE geomembrane/ geosynthetic responses demonstrated a less dramatic shear strength degradation at normal stresses greater than 100 kPa when compared to results of HDPE geomembrane/ geosynthetic interfaces at the same pressures. In fact once maximum shear stress was reached, a trend of none noticeable post-peak strength loss was observed for several LLDPE geomembrane/ geosynthetic interfaces tested. For example, at normal stresses between 200 and 300 kPa, the residual shear strength of the LLDPE geomembrane/ GTB interfaces was only approximately 0.6 to 2.9% lower than its peak shear strength. This behaviour may be attributed to the flexible LLDPE geomembranes experiencing shear failure at an unintended interface. This may indicate that the intended failure surface did not have the lowest shear resistance which could have resulted in the geomembrane experiencing tension near the clamping area of the direct shear box. At high normal stresses, the further increase of the horizontal displacement may have possibly resulted in the elongation of the geomembrane material without loss of strength of the flexible interfaces (Bhatia & Kasturi, 1995).

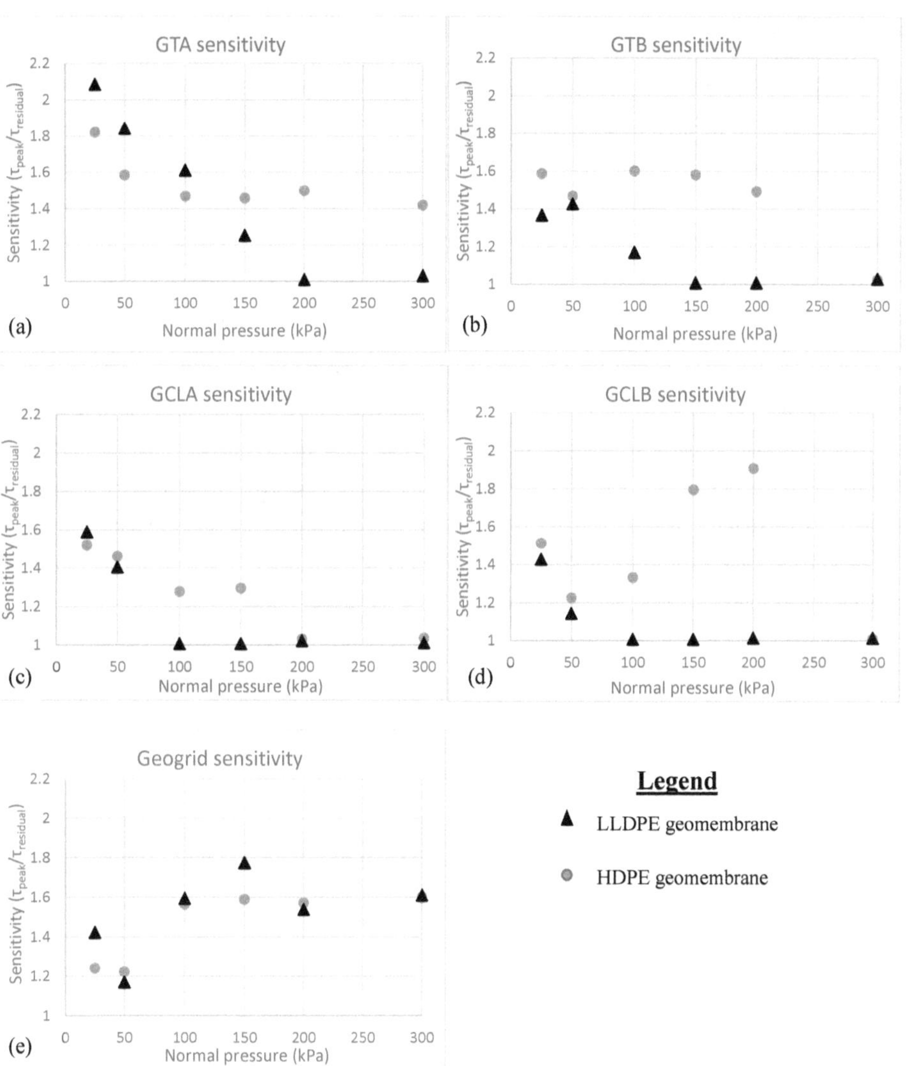

Figure 6-6: Comparison of HDPE and LLDPE geomembrane sensitivity shear strength ratios

6.2.1.4   Residual

After the peak stage, the strain softening phase degrades until a steady residual interface shear strength is reached. From Figure 6-1, it was evident that the residual stress region for majority of the interfaces was indicated by a relatively constant value of shear stress with increasing displacement. The influence of the geomembrane composition on interface residual strength (Table 6-5) was investigated by comparing the magnitudes of residual shear stresses obtained for HDPE and LLDPE geomembrane combinations tested. In Table 6-5, lightly shaded cells indicated the geomembrane with the highest residual value when sheared against the same geosynthetics.

From the table, it was noted that most LLDPE geomembrane interfaces achieved larger shear strength values at normal stresses lower than 150 kPa when sheared against GCL and geotextile interfaces. Once again, this behaviour could be attributed to the interaction mechanisms between the asperities of the flexible LLDPE geomembrane and the adjacent geosynthetic on a superficial level. In the strain softening stage, the stiff HDPE geomembrane asperities tend to pull out and deform more filaments from the adjacent geosynthetic than LLDPE asperities. This results in a lower HDPE/ geomembrane and higher LLDPE/ geomembrane steady state to be achieved during the residual phase.

Table 6-5: Residual shear stresses of geomembrane/ geosynthetic interfaces

	Normal stresses	HDPE geomembrane	LLDPE geomembrane		HDPE geomembrane	LLDPE geomembrane		HDPE geomembrane	LLDPE geomembrane
GTA	25	9.39	12.1	GCLA	19.7	27.9	Geogrid	17	10.2
	50	20.7	22.6		30.2	34.9		21.6	18.9
	100	33.5	39.2		47.4	60.4		25.8	27
	150	44.2	60.7		67.9	71.1		39	26.7
	200	62.3	86.5		98.1	71.9		45.2	50.2
	300	87.3	101		114	88.2		59.1	61.5
GTB	25	12.8	15.9	GCLB	19.1	27.4			
	50	21.6	28.5		31.6	44			
	100	35	62.8		43.9	66.5			
	150	52.8	78.6		40.9	70.9			
	200	64.1	90.2		44.9	72.1			
	300	124	109		111	87.4			

Similar to behaviour experienced at peak stresses, large residual values of HDPE geomembrane interfaces were obtained at normal stresses higher than 200 kPa. This behaviour could be influenced by the large applied confining pressures that caused the HDPE geomembrane asperities and untangled adjacent needle punched geosynthetic surface to embed more tightly together thereby developing an increase in resisting interface shear stresses (Russell, et al., 1998; Kim, 2006). Low resisting stresses were a result of large normal pressures that could compress the flexible LLDPE asperities, which would decrease the ability of interlock between the geosynthetic materials.

In contrast, geogrid interfaces sheared against HDPE and LLDPE geomembranes demonstrated diverse residual stress values. The geogrid residual shear strength revealed no specific trend associated with geomembrane composition. Generally, higher residual values were observed for HDPE geomembranes at low confining pressure (less than 50 kPa) and LLDPE geomembranes at high normal stresses (more than 200 kPa).

Table 6-6 illustrates the percentage difference in the residual shear stresses of the various HDPE and LLDPE geomembrane/ geosynthetic combinations tested at different confining pressures. It can be observed that GTB and the geogrid interfaces reported the maximum and minimum value of variation respectively, compared to other geosynthetic combinations. Furthermore, relatively small residual variances were experienced at high normal pressures (300 kPa) and fairly large residual deviations were observed at high confining stresses (less than 150 kPa). For example, for GTB interfaces, a residual stress difference of 0.44 and 0.12 was obtained for tests conducted at 100 kPa and 300 kPa respectively. This represented 44 % and 12 % difference respectively between the HDPE and LLDPE residual shear stresses.

Table 6-6: Percentage difference of residual shear stresses of various HDPE and LLDPE geomembrane/ geosynthetic interfaces

Normal stresses	GTA	GTB	GCLA	GCLB	Geogrid
25	0.22	0.19	0.29	0.30	0.40
50	0.08	0.24	0.13	0.28	0.13
100	0.15	0.44	0.22	0.34	0.04
150	0.27	0.33	0.05	0.42	0.32
200	0.28	0.29	0.27	0.38	0.10
300	0.14	0.12	0.23	0.21	0.04

Upon comparison of the peak shear stress values (Table 6-4) and residual shear stresses (Table 6-5), it is evident that the LLDPE geomembrane surfaces achieved higher stresses at low normal pressures while HDPE geomembrane combinations obtained larger shear stresses at higher confining pressures for majority GCL and geotextile interfaces.

This indicates that LLDPE geomembrane would produce better shear performance when used in landfill cap applications where low normal pressures are expected. Similarly, HDPE geomembranes would be best for liner applications which would experience high confining pressures such as along landfill bases and side slopes.

6.2.2 Effect of geomembrane composition on vertical displacement

Vertical displacement – time graphs in Figure 6-7 and Figure 6-8 demonstrate the effect of geomembrane composition on vertical displacement (i.e., volume change) for HDPE and LLDPE geomembranes respectively. Both geomembranes were sheared against GTA, GTB, GCLA, GCLB and a geogrid at confining pressures 25 kPa, 50 kPa, 100 kPa, 150 kPa, 200 kPa and 300 kPa.

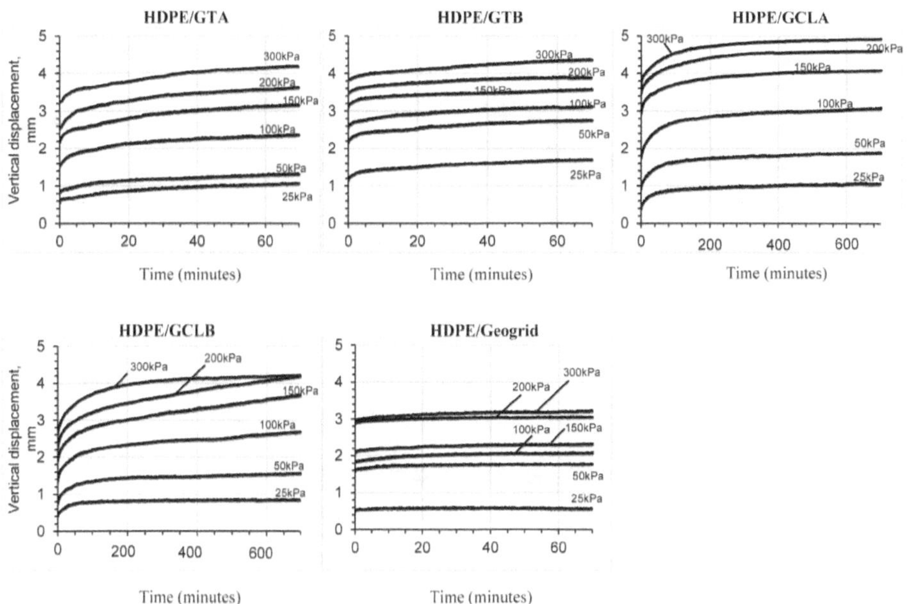

Figure 6-7: Vertical displacement of HDPE geomembrane interfaces sheared against varying geosynthetics

As shown in Figure 6-7 and Figure 6-8, majority of HDPE and LLDPE geomembranes tested against various geosynthetics exhibited larger vertical displacement as the confining pressure increased. Generally most HDPE geomembrane specimen interfaces indicated a continued gradual increase until 70 minutes when sheared against geotextiles and geogrids or 11.7 hours (700 minutes) against GCLs. Majority of LLDPE geomembrane specimens attained vertical displacement equilibrium within 40 minutes or 6 hours (400 minutes) for geotextiles and geogrids, and GCLs respectively. Furthermore, it was observed that the interfaces of GTA and GCLA reported the highest vertical displacement for LLDPE and HDPE geomembranes respectively of 5.3 mm and 4.9 mm respectively.

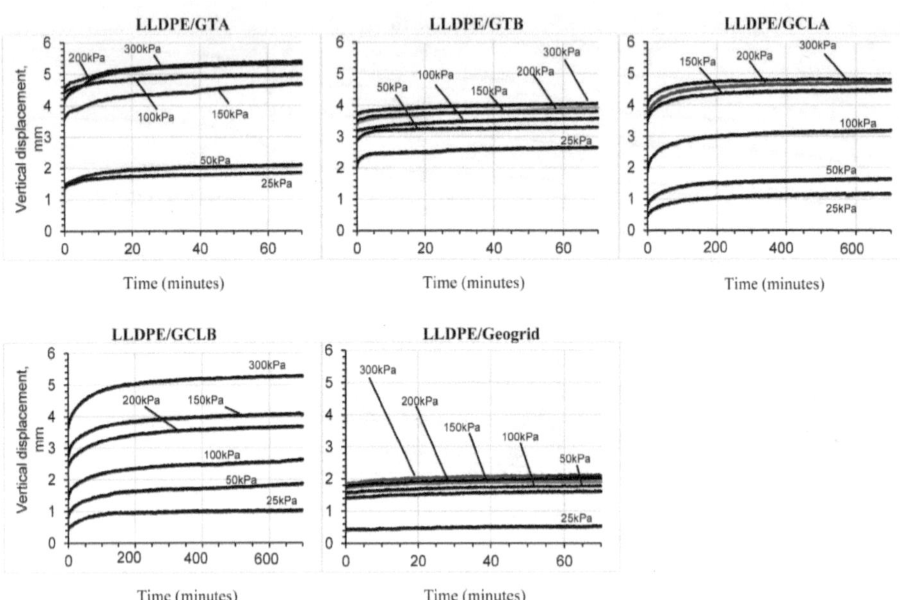

Figure 6-8: Vertical displacement of LLDPE geomembrane interfaces sheared against varying geosynthetics

The varying volume change in the different geosynthetic interfaces may be influenced by the HDPE geomembranes being less likely to allow tight contact between surfaces at low normal stresses.

Consequently, this results in the general HDPE geomembrane interaction improvement and vertical displacement increase during the shearing process.

Table 6-7 and Table 6-8 illustrate the vertical difference (from time 0 minutes to time at end of shearing process) experienced by HDPE and LLDPE geomembrane interfaces at varying confining pressures. It can be observed that GCLs measured the maximum difference compared to the other geosynthetics when sheared against both HDPE and LLDPE geomembranes. This may be caused by normal pressure compression of composite GCLs components and possible extrusion of hydrated bentonite thus enabling the large vertical displacement measurements.

The vertical difference of all LLDPE and HDPE geomembrane interfaces was significantly small. Therefore, application of either geomembrane (on capping or base liners) under expected normal stresses would have almost negligible geomembrane volume influence.

Table 6-7: Influence of confining pressure on the vertical difference of HDPE geomembrane/ geosynthetic interfaces

Confining pressure (kPa)	Vertical difference (mm)				
	Geosynthetics				
	HDPE geomembrane				
	GTA	GTB	GCLA	GCLB	Geogrid
25	0.50	0.57	0.70	0.44	0.10
50	0.51	0.45	0.98	0.80	0.21
100	0.83	0.54	1.33	1.33	0.19
150	0.98	0.59	1.15	1.77	0.25
200	1.09	0.54	1.14	1.95	0.27
300	0.95	0.42	1.22	1.59	0.30

Table 6-8: Influence of confining pressure on the vertical difference of LLDPE geomembrane/ geosynthetic interfaces

Confining pressure (kPa)	Vertical difference (mm)				
	Geosynthetics				
	LLDPE geomembrane				
	GTA	GTB	GCLA	GCLB	Geogrid
25	0.50	0.59	0.73	0.58	0.26
50	0.68	0.44	0.88	1.03	0.30
100	1.31	0.59	1.32	1.23	0.22
150	1.13	0.43	1.10	1.25	0.28
200	0.84	0.69	1.00	1.31	0.26
300	0.56	0.35	0.85	1.62	0.29

6.3 Shear stress – normal stress relationship

In each direct shear test, maximum shear stresses as well as corresponding residual shear stresses were read-off from the graphs in Figure 6-1 and plotted against their respective normal pressures. Best-fit straight lines were fitted through the respective data point results using linear regression analysis methods to form Mohr-Coulomb failure envelopes. The inclination of the failure envelopes to the horizontal axis correspond to the interface angle of shearing resistance between the geomembrane and geosynthetic interface, ϕ'. The intercept on the vertical (shear stress) axis gives the apparent adhesion, denoted by Ca'.

Figure 6-9 to Figure 6-13 present linear (continuous lines) and bilinear (dashed lines) failure envelopes of HDPE and LLDPE geomembranes sheared against other geosynthetics. Each linear failure envelope was characterized using Equation 5-2 to determine ϕ' and Ca' parameters. The bilinear strength envelope was obtained using the same equation fitted to two separate smaller shear stress ranges.

The summary of peak and residual shear strength parameters obtained from various HDPE geomembrane/ geosynthetic and LLDPE geomembrane/ geosynthetic interfaces were quantitatively given in Table 6-9 and Table 6-10 respectively. The indices 'p' and 'r' of the apparent adhesion and friction angle terms stand for peak and residual respectively while 'bp' and 'br' stand for the bilinear relationship at peak and residual shear respectively. The 'bp1' for example, illustrates the first linear line of the peak bilinear relationship for shear at failure while 'bp2' represents the second linear line after the kink in the curve.

The results in Figure 6-9 to Figure 6-13, Table 6-9 and Table 6-10 were used in the discussion of the effects of confining pressure and geomembrane composition on the shear strength characteristics (i.e. friction angle and apparent adhesion). Moreover, the results were utilised to establish how the LLDPE and HDPE geomembranes influenced the interface shear behaviour at high and low normal pressures and whether a systematic difference existed between the two geomembrane friction parameters.

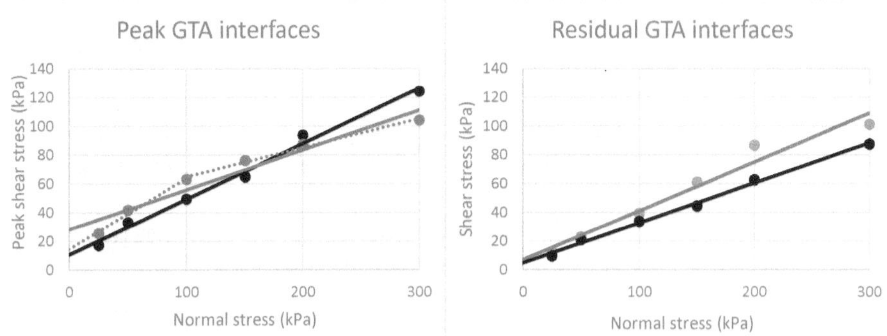

Figure 6-9: Linear and bilinear failure envelopes (at peak and residual) of HDPE and LLDPE geomembranes sheared against GTA material

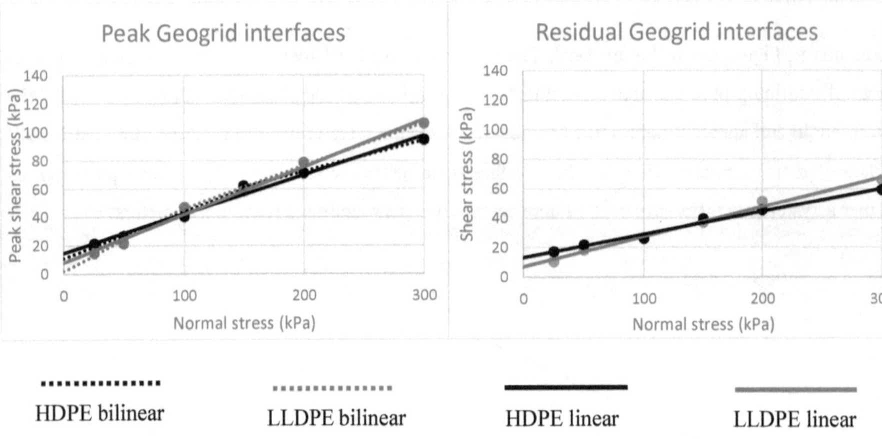

Figure 6-10: Linear and bilinear failure envelopes (at peak and residual) of HDPE and LLDPE geomembranes sheared against geogrid material

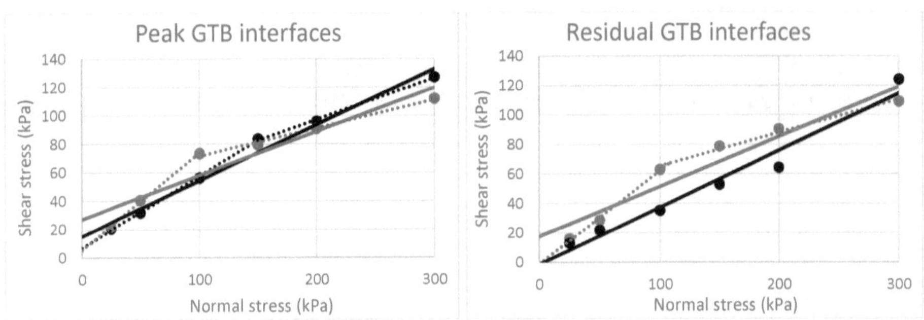

Figure 6-11: Linear and bilinear failure envelopes (at peak and residual) of HDPE and LLDPE geomembranes sheared against GTB material

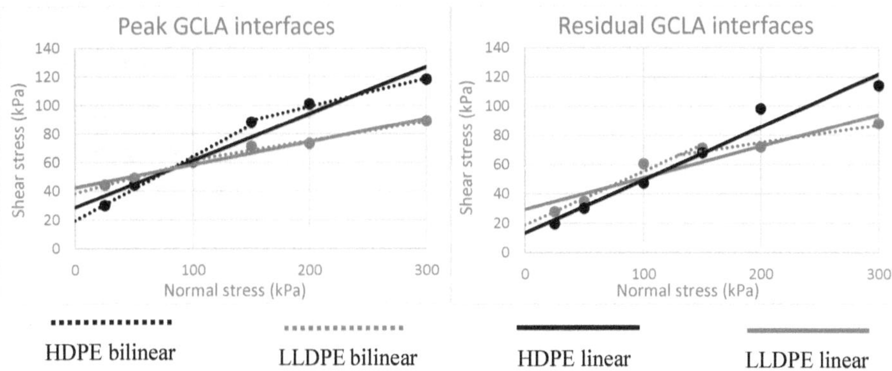

Figure 6-12: Linear and bilinear failure envelopes (at peak and residual) of HDPE and LLDPE geomembranes sheared against GCLA material

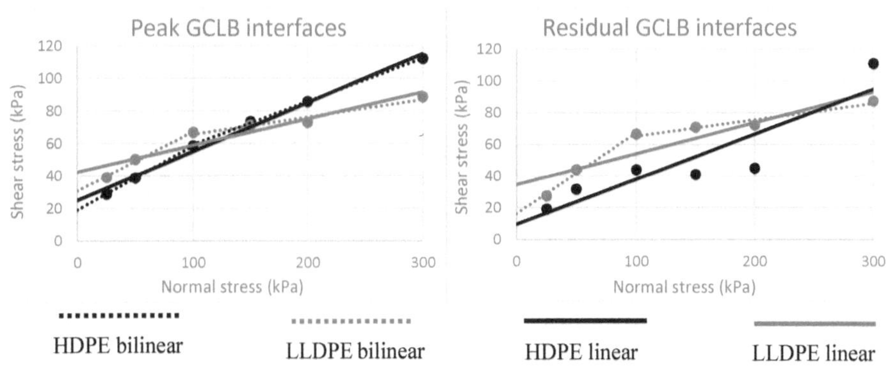

Figure 6-13: Linear and bilinear failure envelopes (at peak and residual) of HDPE and LLDPE geomembranes sheared against GCLB material

Table 6-9: Summary of peak shear strength parameters

Geomembrane	Geosynthetic	Apparent adhesion (kN/m^2)			Friction Angle (degrees)		
		Ca_p'	Ca_{bp1}'	Ca_{bp2}'	ϕ_p'	ϕ_{bp1}'	ϕ_{bp2}'
HDPE	GTA	10.45	-	-	21.11	-	-
	GTB	15.05	6.90	38.37	21.44	26.78	16.37
	GCLA	28.57	19.36	59.81	18.13	24.07	11.09
	GCLB	25.02	19.06	32.70	16.67	21.52	14.84
	Geogrid	14.72	11.00	28.32	15.41	18.07	12.39
LLDPE	GTA	28.12	14.54	44.70	15.47	26.28	11.41
	GTB	27.17	5.41	51.57	17.15	34.36	11.23
	GCLA	42.51	38.484	48.36	9.15	12.44	7.68
	GCLB	42.41	30.86	54.87	9.28	20.03	6.11
	Geogrid	9.68	2.11	15.97	18.03	23.77	16.79

Table 6-10: Summary of residual shear strength parameters

Geomembrane	Geosynthetic	Apparent adhesion (kN/m^2)			Friction Angle (degrees)		
		Ca_r'	Ca_{br1}'	Ca_{br2}'	ϕ_r'	ϕ_{br1}'	ϕ_{br2}'
HDPE	GTA	4.71	-	-	15.53	-	-
	GTB	-1.386	-	-	21.13	-	-
	GCLA	13.21	-	-	19.87	-	-
	GCLB	9.62	-	-	15.82	-	-
	Geogrid	13.17	-	-	8.87	-	-
LLDPE	GTA	6.91	-	-	18.79	-	-
	GTB	17.59	-1.15	42.77	18.72	32.34	12.74
	GCLA	29.43	18.98	50.88	12.17	20.04	6.89
	GCLB	34.62	16.24	54.90	11.02	27.04	5.89
	Geogrid	8.33	-	-	10.91	-	-

Based on the results in Table 6-9 and Table 6-10, the peak ($\phi p'$) and residual ($\phi r'$) friction angles of HDPE geomembranes achieved the highest friction resistance angle. This was true for majority HDPE

geomembrane interfaces excluding geogrid surfaces, which attained angles of interface friction of 15.4°
and 8.9° for peak and residual parameters respectively. LLDPE geomembrane /geogrid surfaces
demonstrated frictional resistance angles of 18° and 10.9° indicating a difference of 14.4 % and 18.7%
between the peak and residual angles of the two polyethylene geomembranes respectively, as illustrated
in

Table 6-11. From the table, it was evident that a considerable difference in friction angle occurred between HDPE
GM/ GCL and LLDPE GM/ GCL interfaces for both peak and residual results.

Table 6-11: Percentage difference between HDPE and LLDPE geomembrane/ geosynthetic linear friction angles

ϕ'	Peak (%)	Residual (%)
GCLA	49.5	38.8
GCLB	44.3	30.3
GTA	26.7	17.3
GTB	20.0	11.4
Geogrid	14.4	18.7

In contrast to the interface angle, the apparent adhesion of peak ($C\alpha_p'$) and residual ($C\alpha_r'$) LLDPE
geomembrane interfaces was significantly higher than that of HDPE geomembranes. Comparably to the
interface friction angle, the apparent adhesion of the HDPE geomembrane/ geogrid was larger than its
LLDPE counterpart. The LLDPE and HDPE geomembranes had a peak apparent adhesion component
of 9.7 kPa and 14.7 kPa respectively when sheared against the geogrid. Based on Table 6-12, the
percentage difference between the HDPE and LLDPE geomembrane/ geogrid apparent adhesion values
at peak was equivalent to 34.2 %. A difference of 36.8 % was calculated at residual apparent adhesion
parameters for LLDPE and HDPE geomembrane/ geogrid interfaces. From the table, it was noticeable
that a significant difference between HDPE and LLDPE geomembrane apparent adhesion values existed
for majority of the tested combinations for both peak and residual results.

It was also clear that both peak and residual HDPE geomembrane interfaces consistently yielded high friction angles with low apparent adhesion values for geotextile and GCL combinations. Moreover, the opposite was true for majority LLDPE geomembrane interfaces, where low friction angles with high apparent adhesion parameters were obtained.

Table 6-12: Percentage difference between HDPE and LLDPE geomembrane/ geosynthetic apparent adhesion values

$C\alpha'$	Peak (%)	Residual (%)
GCLA	32.8	55.1
GCLB	41.0	72.2
GTA	62.8	31.8
GTB	44.6	107.9
Geogrid	34.2	36.8

The peak and residual Mohr-Coulomb failure envelopes in Figure 6-9 to Figure 6-13 and the shear strength characteristics ($C\alpha'$, ϕ') presented in Table 6-9 and Table 6-10 demonstrate the effect of vertical confining pressure on shear strength parameters. From the results, it can be observed that the conventional linear failure envelope was better approximated as a bilinear failure envelope. Experimental observations showed that the shear stress–normal pressure relationship was linear until a critical confining stress was attained forming a bilinear curve. The bilinear failure envelopes gave an abrupt change in friction angle and apparent adhesion at the critcal confining stress as shearing normal stress increased. Pressures lower than the critical confining stress could be governed by low embedment and interlock between the geomembrane/ geosynthetic interfaces. Beyond this confining pressure, it was anticipated the geomembrane asperity deformation and deep interlocking between the contact surfaces controlled failure (Kim, 2006).

The critical confining pressure was found to be in the range of 100 kPa to 150 kPa, similar to findings by Triplett & Fox (2001) and McCartney, et al. (2004). From the results in Table 6-9 and Table 6-10, it was evident that not all interfaces tested expressed a bilinear relationship, other interfaces were simply best deomonstrated as only linear failure envelopes. In this book, approximately 65 % of the tests were better defined as bilinear failure envelopes whereas 35 % were strictly linear envelopes (e.g. HDPE GM/ GTA at peak and residual, LLDPE GM/ GTA at residual, HDPE GM/ GTB at residual, HDPE GM/ GCLA at residual, HDPE GM/ GCLB at residual, HDPE GM/ geogrid at residual and LLDPE GM/ geogrid at residual).

The failure envelope before the critical confining pressure could occur where relatively low normal stresses are experienced (e.g. normal stresses immediately post-construction landfill base or loading conditions experienced by a landfill cover system). The failure envelope after the critical confining pressure could be formed at relatively high normal stresses (e.g. landfill base liner experiencing intermediate or long-term loading of waste). The normal stresses represent the varying loading conditions experienced by a typical landfill base liner system throughout the design life of a containment facility. It was observed in Table 6-9 and Table 6-10 that the bilinear relationship generally demonstrated high frictional resistance angle with low apparent adhesion at normal stresses before the critical confining pressure. A decrease of the friction angle with a significant increase in apparent adhesion was recorded for all results at higher confining stresses, which was consistent with test results reported from previous research (Triplett & Fox, 2001; Thiel, 2001; McCartney, et al., 2004).

A comparison of the linear and bilinear failure envelopes showed that depending on the normal stress range considered, the linear model provided an underestimation or overestimation for shear stresses obtained. Signifying that results presented by the linear relationship over large confining stresses were inaccurate. This emphasised the need to report applicable normal stress ranges with interface strength parameters obtained when using linear equations. Furthermore, it was recognised that a bilinear failure envelope may be more appropriate over large normal stress ranges and at normal stresses which exceed the critical confining stress, as predicted in previous studies (e.g., Triplett & Fox, 2001).

Although the selection to use linear, bilinear or multilinear failure envelops for a design is the decision of the designer(s), identification of the normal stress range expected to be experienced by a specific practical design is recommended.

7. Practical design application

7.1 Introduction

This chapter discussed the application of the results generated from this book and presented practical design examples of a landfill cover and base lining system analysed using limit equilibrium analysis methods. The purpose of this work was to demonstrate the difference in shear strength produced by friction parameters of High Density Polyethylene (HDPE) and Linear Low Density Polyethylene (LLDPE) geomembranes against different geosynthetic combinations. The parameters were selected from tests conducted in the range of confining pressures (25 kPa to 300 kPa) that simulate practical loading conditions in each landfill lining system analysed.

For the purposes of illustration, as well as to avoid repetition, sand was considered as the cover material of the landfill capping structure (illustrated in Figure 3-9). The design principle however, applies equally to capping systems overlaid with other soil types.

7.2 The proposed design methods

The design of a soil structure incorporating geosynthetic material required verification of external and the internal stability. External stability considered the structure to act as a rigid block and was calculated to ensure the bearing capacity was adequate, the rigid block did not slide or overturn and the overall slope stability was sufficient. The internal stability was concerned with the influence of the geosynthetics on the overall structure and the internal stability of each geosynthetic element.

Although both the external and internal stability of soil structures must be considered in the design procedures, this research work focussed on the internal modes of failure. In this book, the application of the external stability primarily addressed the transfer of the surcharges (live and/ or dead loads) at the interfaces, which influenced the bonding of the various geosynthetic layers. The design procedures discussed in this chapter focused on internal stability analysis of the design example, particularly regarding aspects of geosynthetic interface interaction.

The internal stability of structures with geosynthetics could be analysed using a number of limit equilibrium design methods. Limit equilibrium methods could be applied to many geotechnical problems and have been the most widely used analytical technique within the context of slope stability analysis (Duncan, 1996). In principle, the methods are all concerned with satisfying boundary conditions, force and/or moment equilibrium and the failure criterion along an assumed slip surface. This failure surface may be a circular or non-circular arc, a logarithmic spiral or any other arbitrary surface.

In South Africa, the limit equilibrium analysis method for landfill liner stability was typically a preferred approach adopted in design practice (Dookhi, 2013). In this study, the two-part wedge limit equilibrium approach discussed in Chapter 3.5.1.2 was chosen in the design of the forthcoming landfill cover problem. The method provided a reasonable representation of the potential failure surfaces in the design example illustrated in Figure 3-9. Therefore, the finite length slope analysis was carried out to identify the most critical interface in terms of stability as shown in Figure 3-4. The two-part wedge method was based on the computation of the balance of forces acting on a passive wedge at the toe and a long thin active wedge extending the length of the finite slope separated from the remaining cover soil by a tension crack at the crest. The cover soil was limited by an assumed critical failure surface between the geosynthetic and geomembrane interface.

Similarly, a two-part wedge method for translational failure analysis using limit equilibrium method developed by Qian (2008a) was selected for the proposed landfill base design. The base configuration to be analysed was shown in Figure 3-3. The analysis method demonstrated in Chapter 3.5.1.1 illustrated a reasonable replication of the base lining system, therefore the method was applied to calculate the factor of safety (FS) for the waste mass against possible translational failure on predetermined sliding failure surfaces (Qian, 2008a). The waste mass had an active wedge lying on the back slope that was inclined to cause failure and a passive wedge lying on the landfill's foundation soil or base liner system which was likely to resist possible failure.

7.3 Design example

A hypothetical ground excavation with a back slope angle of 18.4 degrees (3H:1V) and waste front slope angle of 14 degrees (4H:1V) (both measured from the horizontal) was proposed for the landfill base design shown in Figure 7-1 and Figure 7-2. With existing foundation conditions deemed adequate for construction, the landfill cell subgrade had an anticipated angle of 1.1 degrees (measured from the horizontal) and a suggested back slope height of 30 m. Waste mass was to be placed in the landfill excavation on top of a multi-layered geosynthetic liner system consisting of geomembranes, GCLs and geotextiles. The waste mass was assumed to have a top width of 20 m, unit weight of 10.2 kN/m^3 and internal friction angle of 33 degrees. The ground water table was assumed to be well below the foundation soil layer and therefore the influence of pore water pressures was neglected.

Figure 7-1: Example of landfill system illustrating the proposed cover slope angle

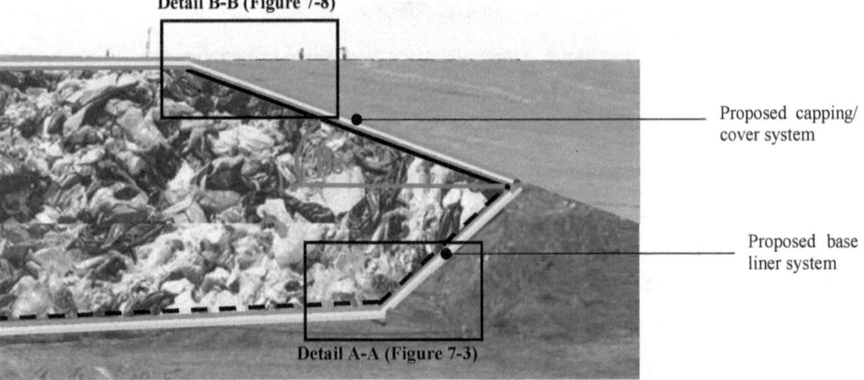

Figure 7-2: Cross section of the landfill system illustrating the proposed cover slope angle

Once the proposed landfill life span was reached, a landfill cover system with a capping soil would be placed over the waste material as shown in Figure 7-2. A uniformly thick 600 mm cover soil was proposed with a slope of 18.4 degrees (3H:1V). A 30 m long slope was suggested with sand as the capping soil having a friction angle of 30 degrees, zero cohesion and a unit weight of 20 kN/m^3. The cover sand was to be placed on a multi-layered geosynthetic liner system (consisting of geomembranes, geotextiles, GCLs and geogrids) where a low shear strength interface would be located beneath an overlying geosynthetic.

The liner interface friction angle and apparent adhesion parameters of the critical interfaces in the landfill base and capping system were assumed to be those listed in Table 7-1. These parameters were obtained from direct shear testing conducted in this study.

The anticipated stress levels in the field guided the selection of linear or bilinear residual shear strength parameters in the proposed design. The base liner would be subjected to lower and higher confining pressures at different stages of construction thus a linear Mohr-Coulomb failure envelope was chosen for this design. The strength envelopes of the bilinear portion of the Mohr-Coulomb failure envelope were more appropriate for the design of the capping systems liners, which would be exposed to mainly low field stresses.

In geotechnical engineering practice, interpretation of adhesion is very project specific. Laboratory test reports involving geosynthetics often indicate a non-zero y-intercept (adhesion) because laboratory interface friction parameters are largely influenced by the testing device (e.g. The Substrate, gripping mechanisms used etc). The ultimate decision whether to include the reported adhesion in a slope stability analysis rests with the design engineer (ASTM D7702). In this design example, the decision to consider the adhesion parameters during the analysis was made.

Table 7-1: Direct shear residual interface friction parameters for critical interfaces

Critical interface	HDPE						LLDPE					
	Linear		Bilinear				Linear		Bilinear			
	$C\alpha_r'$	ϕ_r'	$C\alpha_{br1}'$	ϕ_{br1}'	$C\alpha_{br2}'$	ϕ_{br2}'	$C\alpha_r'$	ϕ_r'	$C\alpha_{br1}'$	ϕ_{br1}'	$C\alpha_{br2}'$	ϕ_{br2}'
GTA	4.7	15.5	-	-	-	-	6.9	18.8	-	-	-	-
GTB	0	21.1	-	-	-	-	17.6	18.7	0	32.3	42.8	12.7
GCLA	13.2	19.9	-	-	-	-	29.4	12.2	19.0	20.0	50.9	6.9
GCLB	9.6	15.8	-	-	-	-	34.6	11.0	16.2	27.0	54.9	5.9
Geogrid	13.2	8.9	-	-	-	-	8.3	10.9	-	-	-	-

In the current design problem, limit equilibrium concepts were used to analyse the landfill and illustrate the factor of safety values produced by the results of the different liner system components. In order to analyse the effects of the liner interface friction angle and apparent adhesion on the landfill base and capping system stability, a comparison of safety factors of HDPE and LLDPE geomembranes against various interfaces of a geosynthetic multilayer liner system was undertaken. The critical interface in the geosynthetic multilayer liner system had to satisfy a minimum FS of 1.3 as recommended by the Department of Water Affairs and Forestry (1998b).

7.4 Design solutions

7.4.1 Base liner system

The base liner system was analysed with a detailed two-part wedge analysis of the base and back slope presented in Figure 7-3. The liner system had base properties highlighted in Section 7.3 and which are also summarised in Table 7-2. According to the analysis procedure, the critical interface obtained the least FS amongst the liner interfaces calculated. Table 7-4 shows that the minimum FS were 1.16 and 1.47 for GTA sheared against HDPE and LLDPE geomembranes respectively. The table indicates that the least FS would occur at the same interface for both geomembranes considered experiencing similar conditions.

Table 7-4 shows that the least FS calculated (1.16) for the HDPE geomembrane-GTA interface would be below the specified minimum FS recommended by the Department of Water Affairs and Forestry (1998b) thus considered unacceptable.

Design properties were implemented into Equation 3-4 and Equation 3-5 which allowed WA, WP and WT parameters (highlighted in Table 7-3) to be calculated. Ultimately, these parameters were used in Equation 3-1 to calculate FS values produced in Table 7-4. All detailed calculations are attached in Appendix VI.

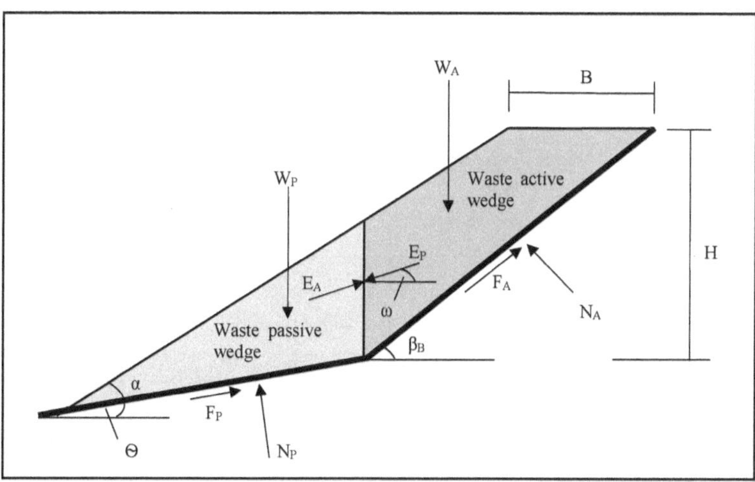

Figure 7-3: Two-part wedge limit equilibrium forces for waste mass over multi-layered geosynthetics (**Detail A-A**) (Qian & Koerner, 2004)

Table 7-2: Proposed base liner system dimensions and properties

Property	Abbreviation	Value	Units
Top width of waste mass	B	20	m
Height of back slope	H	30	m
Angle of front slope	α	14	degrees
Angle of back slope	β_B	18.4	degrees
Unit weight of solid waste	γ_{sw}	10.2	kN/m^3
Angle of landfill cell subgrade	θ	1.1	degrees
Internal friction angle of solid waste	ϕ_{sw}	33	degrees

Table 7-3: Factor of safety parameters W_A, W_P and W_T

Property	Measure	Units
W_A	13795.1	N/A
W_P	3463.5	N/A
W_T	17258.6	N/A

Table 7-4: Factors of safety for each interface without consideration of leachate level

Critical interface	HDPE			LLDPE		
	ϕ_r'	$C\alpha_r'$	FS	ϕ_r'	$C\alpha_r'$	FS
GTA	15.5	4.7	1.16	18.8	6.9	1.47
GTB	21.1	0	1.40	18.7	17.6	1.83
GCLA	19.9	13.2	1.77	12.2	29.4	1.78
GCLB	18.8	9.6	1.57	11	34.6	1.87

In the analysis, it was assumed that the leachate level was zero and the potential translational failure surface in the liner system would pass through the same interface at both the back slope and base. However, it was common that the critical liner interface may occur within one interface at the back slope and another interface at the base as indicated by Qian (2008b).

Calculating FS along the same interface at the back slope and base may result in unsafe FS values (Qian, 2008a). In order to determine whether failure would occur at the same geomembrane/ GTA interface for

HDPE and LLDPE geomembranes, different back slope and base liner combinations were considered. The FS values are illustrated in Table 7-5, which shows the critical interfaces where failure was anticipated to occur.

Table 7-5: Factors of safety for various interface combinations without consideration of leachate level

HDPE		Active wedge			
		GTA	GTB	GCLA	GCLB
Passive wedge	GTA	1.19	1.41	1.62	1.48
	GTB	1.23	1.45	1.66	1.52
	GCLA	1.35	1.57	1.78	1.64
	GCLB	1.30	1.52	1.73	1.59
LLDPE		Active wedge			
		GTA	GTB	GCLA	GCLB
Passive wedge	GTA	1.50	1.72	1.61	1.65
	GTB	1.61	1.84	1.73	1.77
	GCLA	1.64	1.87	1.76	1.81
	GCLB	1.68	1.91	1.81	1.85

Once again, the critical interfaces which presented the minimum FS were HDPE geomembrane/ GTA (FS = 1.19) and LLDPE geomembrane/ GTA (FS = 1.50) interfaces, when the leachate level was assumed to be zero. Thus, least FS would occur on the back and base slopes on the GTA interfaces for both HDPE and LLDPE geomembranes. The analysis indicated HDPE and LLDPE geomembranes showed parallel performance patterns although producing different FS values.

Table 7-5 shows that the FS values calculated for the HDPE geomembrane-GTA and HDPE geomembrane-GTB interfaces would be below the specified minimum FS recommended by the Department of Water Affairs and Forestry (1998b) thus considered unacceptable.

7.4.1.1   Effect of varying waste depth

The effect of waste filling on the critical interface was demonstrated by detailed values of FS calculated from various combinations of widths of waste mass (B) and heights of the back slope (H) illustrated in

Figure 7-4 and listed in Table 7-6 and Table 7-7. The bases for the various combinations was from typical waste filling procedures.

From Table 7-6 and Table 7-7, the critical interface was located at the GTA or GTB interface of the subgrade and base slope of HDPE geomembrane surfaces at various waste filling conditions. On the other hand, LLDPE geomembrane interfaces experienced significantly varied critical interfaces for the varying waste depths.

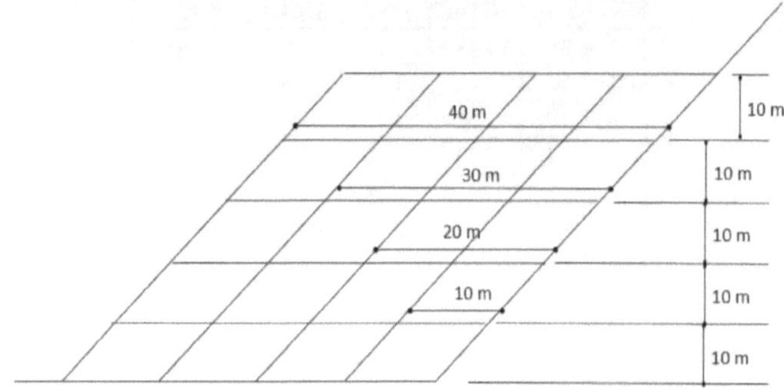

Figure 7-4: Example of possible waste filling procedure

It was noted, the factor of safety obtained for LLDPE geomembrane interfaces exceeded that of HDPE geomembrane surfaces in majority of the investigations. Once again, this behaviour could be a result of the interaction of the contact surfaces mainly between the opposite geosynthetic planes and the asperities of the geomembranes. The flexible LLDPE geomembrane surfaces would be able to adjust and fit closely to the adjacent geosynthetic shape, which enabled the asperities to embed and interlock tightly within the adjacent geosynthetic surface thus increasing the shear stress and FS achieved.

In the analyses where the waste depth changed, more HDPE geomembrane interfaces did not achieve FS values greater than 1.3 which was recommended by the Department of Water Affairs and Forestry (1998b) for a safe slope design, than LLDPE geomembrane interfaces. Thus it was anticipated that the application of LLDPE geomembranes (obtained minimum FS value = 1.22) at a base lining system would produce effective stability in comparison to HDPE geomembranes (obtained minimum FS value = 1.03).

However, HDPE geomembrane interfaces would be most viable for base lining systems. The effectivness of these geomembranes has been demostrated in Section 7.4.1.1 indicating that critical interfaces of HDPE geomembrane combinations vary less when compared to those of LLDPE geomembranes, hence making the predetermined critical interface easier to locate during design.

Table 7-6: Changes in critical interface and factor of safety with varying waste depth for HDPE geomembrane interfaces

H (m)	B (m)															
	10		20		30		40		50		60		70		80	
	FS	Interface	FS	Interface	FS	Interface	FS	Interface	FS	Interface	FS	Interface	FS	Interface	FS	Interface
10	1.548	GTA/GTB	2.022	GTA/GTB	2.548	GTA/GTA	3.135	GTA/GTA	3.767	GTA/GTA	4.425	GTA/GTA	5.087	GTA/GTA	5.740	GTA/GTA
20	1.222	GTA/GTA	1.369	GTA/GTA	1.547	GTA/GTA	1.752	GTA/GTA	1.982	GTA/GTA	2.233	GTA/GTA	2.503	GTA/GTA	2.789	GTA/GTA
30	1.110	GTA/GTA	1.192	GTA/GTA	1.290	GTA/GTA	1.401	GTA/GTA	1.524	GTA/GTA	1.660	GTA/GTA	1.806	GTA/GTA	1.963	GTA/GTA
40	1.057	GTA/GTA	1.113	GTA/GTA	1.178	GTA/GTA	1.250	GTA/GTA	1.330	GTA/GTA	1.418	GTA/GTA	1.513	GTA/GTA	1.614	GTA/GTA
50	1.026	GTA/GTA	1.068	GTA/GTA	1.116	GTA/GTA	1.169	GTA/GTA	1.226	GTA/GTA	1.289	GTA/GTA	1.357	GTA/GTA	1.429	GTA/GTA

Table 7-7: Changes in critical interface and factor of safety with varying waste depth for LLDPE geomembrane interfaces

H (m)	B (m)															
	10		20		30		40		50		60		70		80	
	FS	Interface	FS	Interface	FS	Interface	FS	Interface	FS	Interface	FS	Interface	FS	Interface	FS	Interface
10	1.982	GTA/GTA	2.353	GTA/GTA	2.779	GTA/GTA	3.240	GTA/GTA	3.721	GTA/GTA	4.206	GTA/GTA	4.683	GTA/GTA	5.143	GTA/GTA
20	1.476	GTA/GTA	1.598	GTA/GTA	1.740	GTA/GTA	1.899	GTA/GTA	2.074	GTA/GTA	2.262	GTA/GTA	2.461	GTA/GTA	2.670	GTA/GTA
30	1.330	GTA/GTA	1.396	GTA/GTA	1.472	GTA/GTA	1.556	GTA/GTA	1.649	GTA/GTA	1.750	GTA/GTA	1.857	GTA/GTA	1.972	GTA/GTA
40	1.261	GCLA/GTA	1.305	GCLA/GTA	1.354	GCLA/GTA	1.409	GCLA/GTA	1.468	GCLA/GTA	1.533	GCLA/GTA	1.602	GCLA/GTA	1.675	GCLA/GTA
50	1.222	GCLA/GTA	1.253	GCLA/GTA	1.289	GCLA/GTA	1.328	GCLA/GTA	1.371	GCLA/GTA	1.417	GCLA/GTA	1.466	GCLA/GTA	1.518	GCLA/GTA

7.4.2 Capping system

The influence of critical interfaces formed by different liner cover system components was investigated by comparing the magnitudes of factor of safety values calculated using the limit equilibrium analysis method illustrated in Figure 7-5. The critical interface in the geosynthetic multilayer liner system had to satisfy a minimum FS of 1.3 as recommended by the Department of Water Affairs and Forestry (1998b).

FS values of potential planes of failure between geomembrane and geosynthetic combinations computed from linear, bilinear (at low normal stresses) and bilinear (at high normal stresses) failure envelopes are shown in Table 7-8.

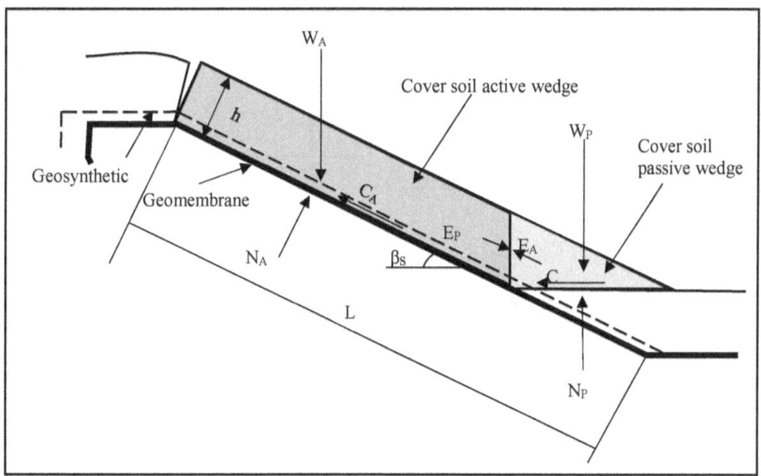

Figure 7-5: Two-part wedge limit equilibrium forces involved in a finite length slope analysis for a uniformly thick cover soil over multi-layered geosynthetics (**Detail B-B**) (Koerner & Soong, 1998)

Minimum FS values of 1.24 and 2.92 for HDPE and LLDPE geomembranes respectively against GTB and GTA respectively, were obtained when a linear envelope was applied for the anticipated critical surfaces. Low FS values were noted for bilinear failure envelopes at low normal stresses while bilinear failure envelopes at high normal stresses showed significantly high FS values for LLDPE geomembrane combinations.

A general observation of the results indicated that the FS values were strongly influenced by the apparent adhesion of the critical interface. Limit equilibrium calculations using large apparent adhesion results obtained high FS values while the use of large interface friction angles had a less dramatic influence on the FS values achieved.

Table 7-8: Factor of safety values for various critical interfaces in the proposed cover design

Critical interface	HDPE			LLDPE		
FS	Linear	Bilinear		Linear	Bilinear	
GTA	2.155	-	-	2.922	-	-
GTB	1.241	-	-	5.740	1.976	12.060
GCLA	4.650	-	-	8.501	6.185	13.889
GCLB	3.467	-	-	9.811	5.897	14.899
Geogrid	4.913	-	-	3.486	-	-

Table 7-9: Proposed cover liner system dimensions and properties

Property	Abbreviations	Value	Units
Unit weight of the cover soil	γ	20	kN/m^3
Thickness of the cover soil	h	0.6	m
Length of slope measured along the geomembrane	L	30	m
Soil slope angle beneath the geomembrane	βs	18.4	degrees
Internal friction angle of the cover soil	φ	30	degrees

Design dimensions and properties in Table 7-9 were implemented into Equation 3-11, Equation 3-12, Equation 3-13 and Equation 3-14 which allowed W_A, N_A, C_a and W_P parameters (highlighted in Appendix VII) to be calculated. Ultimately, these parameters were used in Equation 3-10 to calculate FS values produced in Table 7-8. All detailed calculations are attached in Appendix VII.

From Table 7-8, it was evident that a large number of critical interfaces achieved the recommended FS value. Although the FS obtained from linear, and low and high bilinear parameters were within the recommended safety factor, not all these parameters were an accurate representation of a landfill cover system in which strictly low normal stresses could be experienced. Therefore, FS values obtained from only low normal stresses in the bilinear relationship were considered appropriate for this design analysis. This was mainly because the cover loads in the landfill cap (mostly generated from the cover soil and geosynthetic layers) were expected to produce normal stresses less than 100 kPa (Bacas, et al., 2015), which fell within the range of the bilinear low normal pressures tested.

The application illustrated in this example indicated that sufficient slope stability was attained when both HDPE and LLDPE geomembranes were used in a landfill capping system. Clearly in majority of the tested combinations, the use of LLDPE geomembranes produced significantly better FS values compared to HDPE geomembranes for linear failure envelope parameters when sheared against geotextiles and GCLs. Furthermore, design appropriate bilinear failure envelopes at low normal stresses (for LLDPE geomembrane combinations) appeared to achieve FS values above the minimum recommended by the Department of Water Affairs and Forestry (1998b). Indicating that LLDPE geomembranes could be assumed best suited for landfill capping systems when low normal stress would be experienced on a slope, yet laboratory tests on site-specific construction materials are strongly recommended before construction commences.

8. Conclusions and recommendations

Large direct shear tests were conducted on HDPE and LLDPE geomembranes sheared against geotextiles, GCLs and a geogrid. These geosynthetic combinations were used to examine the influence of the two polyethylene geomembranes on the stress-displacement relationship and shear strength characteristics. The summary of findings emerging from this study and some recommendations for further research work are presented in the following sections.

8.1 Summary of conclusions

The following conclusions were drawn based on the results and interpretation presented in this research:

1. The shear stress and horizontal displacement relationship of HDPE and LLDPE geomembrane interfaces, in the pre-peak stage, could be represented using a mathematical model. The model provided an accurate approximation of the interface behaviour at constant normal stress by illustrating the initial shear modulus (Ksi) and tangent shear modulus (Kst) of the sheared interface. From the results, there seemed to be an evident trend demonstrating that at higher normal stresses (greater than 200 kPa) the Ksi values for LLDPE geomembrane interfaces become close or higher to the Ksi values of HDPE geomembrane interfaces. This was true for majority of the geosynthetic interactions. This behaviour could be influenced by the large applied confining pressures that caused the flexible LLDPE geomembrane asperities to tightly interact with the adjacent geosynthetic thereby developing increasing resisting interface shear stresses during initial shear. The HDPE geomembrane combinations showed larger Kst values when compared to LLDPE geomembranes tested against majority of geosynthetics indicating larger stiffness and rigidity of HDPE geomembrane interfaces,

2. LLDPE geomembrane/ geosynthetic interfaces presented larger peak and residual shear stress values than HDPE geomembranes sheared against majority of the geosynthetics at normal confining pressures lower than 100 kPa. For normal stresses higher than 100 kPa, large peak and residual values of HDPE geomembrane interfaces were obtained. This was true for interfaces sheared against geotextiles and GCLs. This indicates that LLDPE geomembrane would produce better shear

performance when used in landfill cap applications where low normal pressures are expected. Similarly, this reinforced that HDPE geomembranes would be best for liner applications which would experience high confining pressures such as along landfill bases and side slopes,

3. Both peak and residual linear failure envelopes showed that tough HDPE geomembrane interfaces consistently yielded high friction angles with low apparent adhesion values for majority of the geosynthetic combinations. Moreover, the opposite was true for the softer LLDPE geomembrane interfaces where high apparent adhesion was achieved with low interface friction angles,

4. Due to the effect of normal stress on interface properties, it was observed that the conventional linear failure envelope was better approximated as a bilinear model for both HDPE and LLDPE geomembrane interfaces. Experimental observations showed that the shear stress–normal pressure relationship was linear until a critical confining stress in the range of 100 kPa to 150 kPa was attained, similar to findings by Triplett & Fox (2001) and McCartney, et al. (2004). The bilinear relationship demonstrated an increase in the frictional resistance angle with low apparent adhesion at normal stresses before the critical confining pressure. A decrease of the friction angle with an increase in apparent adhesion occurred for all results at higher confining stresses which was consistent with test results reported from previous research (Triplett & Fox, 2001; Thiel, 2001; McCartney, et al., 2004).

5. The friction parameters generated from this study were applied to a practical design example of a landfill base lining system. Factor of safety results demonstrated that at zero leachate levels, the lowest FS values occur at the same geomembrane/ GTA interface for the cell subgrade and back slopes for both HDPE and LLDPE geomembranes,

6. The effect of various waste filling procedures on the critical interface demonstrated high factor of safety values for majority of the base liner LLDPE geomembrane/ geosynthetic combinations. The results indicate that the application of LLDPE geomembranes at a base lining system would produce effective stability in comparison to HDPE geomembranes at varying waste depths,

7. The application illustrated in the capping system example indicated sufficient slope stability was attained when both HDPE and LLDPE geomembranes were used. In majority of the tested combinations, the use of LLDPE geomembranes produced significantly better factor of safety values compared to HDPE geomembranes for linear and bilinear failure envelope parameters when sheared

against geotextiles and GCLs. Indicating that LLDPE geomembranes would be best suited for landfill capping systems and

8. The comparison of HDPE and LLDPE geomembrane friction parameters revealed different geomembrane performance patterns for different landfill applications. Consequently, benefits associated with various geomembrane types should be considered carefully before construction commences.

8.2 Recommendations

The following are the recommendations for future research:

- It was recommended that an examination of the shear failure experienced between the test specimen and the shearing substrate should be undertaken. Various gripping methods should be analysed to estimate the effect of possible unintended failure has on the shear-displacement curve and shear-normal stress relationship,

- Further investigation was required to understand the effect of stretched geomembrane material (or the adjacent strained geotextile, GCL, geogrid etc) has on interface shear strength parameters obtained,

- Additional research into finite element slope stability analysis exploring interface friction values of geomembranes with various composition should be investigated. Furthermore, this computer software could provide a refined analysis identifying which factors influence the shear stress – horizontal displacement behaviour of geomembrane/ geosynthetic interfaces,

- The use of high digital imaging was suggested in order to be able to identify how geomembrane asperity deformation and fibre reorientation occurred. Visualising the geosynthetic interactions at a microscopic scale could lead to further understanding of interface behaviour,

- The study examined the effect of geomembrane composition at the early stage of a project. Further assessment into the durability and long term interface behaviour of various geomembranes once they had been in contact with leachates was required. Record of possible change in interface friction parameters could determine deformations and mobilised shear resistance in existing waste containment facilities with multi-layered geosynthetics,

- The current study used co-extruded geomembranes from the same manufacturer. An additional study including geomembranes with smooth or different texturing types and possibly from different manufacturers would be essential and

- It was recommended that wherever possible, designers should conduct material-specific testing of interface frictional strength to verify that the materials specified and/or supplied for a project meet the required design requirements, without resorting to the use of generalised friction values for similar geosynthetic interfaces from published data.

References

Agru, 2016. *HDPE and LLDPE Agru smooth liner*. [Online]
Available at: http://agruamerica.com/products/geomembranes/smooth-liner/ [Accessed 12 April 2017].

Allen, J. M. & Fox, P. J., 2007. *Pyramid-tooth gripping surface for GCL shear testing*. North America, Industiral Fabrics Association International.

Anubhav & Basudhar, P. K., 2010. Modeling of soil-woven geotextile interface behavor from direct shear test results. *Geotextiles and geomembranes,* 28(4), pp. 403-408.

Arrighi, V. & Kraft, A., 2011. *Polymer structure and architecture*. [Online]
Available at:
http://www.che.hw.ac.uk/teaching/B11MS1/Material/Week%204/lecture_11PolymerStructureandArchi tecture.htm [Accessed 15 June 2017].

Arulrajah, Arul; Horpibulsuk, Suksun, Maghoolpilehrood, Farshid; Samingthong, Wisanukorn, Du, Yan-Jun & Shen, Shui-Long, 2014a. Evaluation of interface shear strength properties of geogrid reinforced foamed recycled glass using a large scale direct shear testing apparatus. *Journal of materials in civil engineering,* 26(5).

Arulrajah, A., Rahman, M. A., Piratheepan, J. & Bo, M. W., 2014b. Evaluation of interface shear strength properties of geogrid-reinforced construction and demolotion materials using a modified large scale direct shear testing apparatus. *Journal of materials in civil engineering,* 26(5).

ASTM D4439, 2017. *Terminology for geosynthetics,* (ASTM D4439) West Conshohocken: ASTM International.

ASTM D5321, 2008. *Standard test method for determining the coefficient of soil and geosynthetic or geosynthetic and geosynthetic friction by the direct shear method,* (ASTM D5321_08) West Conshohocken: ASTM International.

ASTM D6243, 2009. *Standard test method for determining the internal and interface shear resistance of geosynthetic clay liner by the direct shear method,* (ASTM D6243_09) West Conshokocken: ASTM International.

ASTM D7702, 2014. *Standard guide for considerations when evaluating direct shear results involving geosynthetics,* (ASTM D7702/D7702M_14) West Conshohocken: ASTM International.

Athanasopoulos, G., Vlachakis, V., Zekkos, D. & Spiliotopoulos, G., 2013. *The December 29th 2010 Xerolakka Municipal Soilid Waste landfill failure.* Paris, The 18th International Conference on Soil Mechanics and Geotechnical Engineering.

Bacas, B., Canizal, J. & Konietzky, H., 2015. Shear strength behaviour of geotextile/geomembrane interfaces. *Journal of rock mechanics and geotechnical engineering,* 8(001), pp. 1-8.

Bauer, J., Koelsch, F. & Borgatto, A., 2008. *Stability analysis according to different shear strength concepts exemplified by two case studies.* Hohhaido, Japan, The 5th Asian-Pacific Landfil Symposium.

Bergado, D., Glawe, U., Sia, A.H.I, Youwai, S., & Voottipreux, P., 1997. *Interface shear strengths of different geosynthetics in landfill liner system: A case of Sekaew landfill, Thailand,* Bangkok: Asian Institute of Technology.

Bhatia, S. & Kasturi, G., 1995. *Comparison of PVC and HDPE geomembranes (Interface friction performance),* New York: PVC Geomembrane Institute.

Blond, E. & Elie, G., 2006. *Interface shear-strength properties of textured polyethylene geomembranes,* Canada: Solmax International.

CETCO, 2013. *Sodium bentonite: Its structure and properties,* North America: CETCO.

CETCO, 2014. *Evaluating GCL chemical compatibility,* North America: CETCO.

Clough, G. W. & Duncan, J. M., 1971. Finite element analysis of retaining wall behaviour. *Journal of soil mechanics and foundation engineering,* 97(12), pp. 1657-1672.

Cochran, W. G., 1963. *Sampling Techniques.* 2nd ed. New York: John Wiley and Sons.

Contain Enviro Services, 2016. *HDPE vs LLDPE geomembrane liners.* [Online]
Available at: https://www.linkedin.com/pulse/hdpe-vs-lldpe-geomembrane-liners-contain-enviro-services [Accessed 12 April 2017].

Costley, S., 2013. *Waste Classification and Management Regulations and Supporting Norms & Standards.* [Online] Available at:
https://www.environment.gov.za/sites/default/files/docs/wasteclassification_regulations_disposalstandards.pdf [Accessed 9 April 2015].

Davison, L. & Springman, S., 2000. *Basic mechanics of soils.* [Online]
Available at: http://environment.uwe.ac.uk/geocal/SoilMech/basic/soilbasi.htm [Accessed 4 August 2017].

Department of Environmental Affairs, 2013a. *Waste classification and management regulations,* South Africa: Government Gazette No. 36784, Government Notice No. R634.

Department of Environmental Affairs, 2013b. *National norms and standards for disposal of waste to landfill,* South Africa: Government gazette No.36784, Government Notice No. R636.

Department of Environmental Affairs, 2013c. *National norms and sandards for the storage of waste,* South Africa: Government Gazette No.37088, Government Notice No.926 .

Department of Environmental Affairs, 2016. *SAWIC implementation.* [Online]
Available at: http://sawic.environment.gov.za/?menu=77 [Accessed 31 May 2017].

Department of Water Affairs and Forestry, 1996. *South african water quality guidelines,* Pretoria: Department of water affairs and forestry.

Department of Water Affairs and Forestry, 1998a. *Minimum requirements for the handling, classification and disposal of hazardous waste,* South Africa: Second Edition 1998.

Department of Water Affairs and Forestry, 1998b. *Minimum requirements for waste disposal by landfill: Second Edition 1998,* Pretoria: Department of Water Affairs and Forestry.

Department of Water Affairs and Forestry, 1998c. *Waste disposal on land.* [Online]
Available at: https://www.dwa.gov.za/dir_wqm/wqm_wasteDispLand.htm [Accessed 6 June 2016].

Dookhi, A. S., 2013. *The lining of steep landfill slopes in South Africa and the applicability of the "Minimum requirments for waste disposal by landfill" by the Department of Water Affairs and Forestry,* Masters book, Durban: University of KwaZulu-Natal.

Duffy, D. P., 2016. *Three types of slope failure in landfills.* [Online]
Available at: http://foresternetwork.com/daily/waste/landfill-management/three-types-of-failures-in-landfills/ [Accessed 16 June 2016].

Duncan, J. M., 1996. State of the art: Limit equilibrium and finite element analysis of slopes. *Journal of geotechnical engineering,* 122(7), pp. 577-596.

Emery, R., 2014. *New landfill classifications,* CIV5124Z - University of Cape Town Geosynthetic Course Lecture Notes, Cape Town: Jeffares & Green.

E-Square engineering, 2015. *Detailed design report landfill liner designs,* Pretoria: Transnet.

Esterhuizen, J. J. B., Filz, G. M. & Duncan, J. M., 2001. Constitutive behaviour of geosynthetic interfaces. *Journal of geotechnical and geoenvironmental engineering,* 127(10), pp. 834-840.

Feng, S. J., Chen, Y. M., Gao, L. Y. & Geo, G. Y., 2010. Translational failure analysis of landfill with retaining wall along the underlying liner system. *Environment Earth Science,* Volume 60, pp. 21-34.

Feng, S. & Lu, S., 2016. Repeated shear behaviors of geotextile/ geomembrane and geomembrane/ clay interfaces. *Environmental Earth Sciences,* 75(273).

Fibromat, 2014. *Geosynthetic clay liner (GCL) - the optimum lining solution.* [Online]
Available at: http://www.fibromat.com.my/products/geosynthetic-clay-liner/ [Accessed 14 July 2016].

Fowmes, G. J., Dixon, N. & Jones, D. R. V., 2008. Validation of a numerical modelling technique for multilayered geosynthetic landfill lining systems. *Geotextiles and geomembranes,* 26(2), pp. 109-121.

Fox, P. J. & Ross, J. D., 2011. Relationship between NP GCL internal and HDPE GMX/ NP GCL interface shear strengths. *Journal of geotechnical and geoenvironmental engineering,* 137(8), pp. 743-753.

Fox, P. J. & Stark, T. D., 2015. State-of-the-art report: GCL shear strength and its measurement - ten-year update. *Geosynthetics International,* 22(1), pp. 3-47.

Fox, P. & Kim, R., 2008. Effect of progressive failure on measured shear strength of geomembrane/ GCL interface. *Journal of getotechnical and geoenvironmental engineering,* 134(4), pp. 459-469.

Fox, P., Morrison, T., Nye, Christopher; Hunter, Jay & Olsta, James, 2014. *Current research on dynamic shear behaviour of needle-punched geosynthetic clay liners,* Columbus, OH USA: CETCO.

Fox, P. & Stark, T., 2004. State-of-the-art report: GCL shear strength and its measurement. *Geosynthetics International,* 3(11), pp. 141-175.

Foye, K., 2011. Armored geomembrane cover engineering. *International journal of environmental research and public health,* Volume 8, pp. 2240-2264.

Geocomp corporation, 2012. *Control and report software for fully automated direct shear tests on ShearTrac-III systems using windowsXP/Vista/7,* USA: ShearTrac-III.

Geofabrics, 2001. *Shear box testing summary: geosynthetic materials intended for use in geomembrane protection applications,* Stourton: Geofabrics.

Geofabrics, 2015. *Bidim - Nonwoven geotextiles.* [Online]
Available at: http://www.geofabrics.com.au/products/products/1-bidim-nonwoven-geotextiles/overview [Accessed 2 November 2015].

Geofabrics, 2016. *Texcal Nonwoven staple fibre geotextiles.* [Online]
Available at: http://www.geofabrics.com.au/products/products/48-texcel-nonwoven-staple-fibre-geotextiles/functions-and-applications [Accessed 5 March 2017].

Geosynthetic Institute, 2012. *Determining the long-term strength of flexible geogrids,* Folsom: Geosynthetic Institute. GRI Standard Practice GG4(b).

GIGSA, 2009. *Environmental protection including waste containment.* [Online]
Available at: http://www.gigsa.org/Newsletters/GIGSA_November2009.pdf [Accessed 19 January 2017].

Gilbert, R. B. & Byrne, R. J., 1996. Strain-softening behaviour of waste containment system interfaces. *Geosynthetics International,* 3(2), pp. 181-203.

Giroud, J. P. & Beech, J. F., 1989. *Stability of soil layers on geosynthetic lining systems.* St Paul, Proceeding geosynthetics 1989.

Global plastic sheeting, 2017a. *HDPE, LDPE, LLDPE - What's the difference anyway?.* [Online]
Available at: https://www.globalplasticsheeting.com/hdpe-vs-lldpe-vs-ldpe [Accessed 11 April 2017].

Global plastic sheeting, 2017b. *LLDPE - Linear low density polyethylene.* [Online]
Available at: https://www.globalplasticsheeting.com/lldpe-linear-low-density-polyethylene-liners [Accessed 11 April 2017].

Gomez, J. E., Filz, G. M. & Ebling, R. M., 2000. *Development of an improved numerical model for concrete to soil interfaces in soil-structure interaction analyses,* U.S Army Corps of Engineers, Engineer Research and Development Center: Information technology laboratory technical report. ITL-99-1.

GSE Environmental, 2015. *GSE HD textured geomembrane.* [Online]
Available at:
http://www.gseworld.com/content/documents/datasheets/membranes/Europe_and_Africa/HD_Textured_-_ISO_-_EuropeAfrica.pdf [Accessed 17 May 2015].

Hebeler, G. L., Frost, J. D. & Myers, A. T., 2005. Quantifying hook and loop interaction in textured geomembrane-geotextile systems. *Geotextile and geomembrane,* 23(1), pp. 77-105.

Hillman, R. P. & Stark, T. D., 2001. *Shear strength characteristics of PVC geomembrane-geosynthetic interfaces.* California, Geosynthetics International, Vol. 8, No. 2, pp. 135-162.

Hossain, B., Hossain, Z. & Sakai, T., 2012. Interaction properties of geosynthetic with different backfill soils. *International journal of geoscience,* Volume 3, pp. 1033-1039.

Hungr, O. & Morgenstern, N. R., 1984. High velocity ring shear tests on sand. *Geotechnique,* 34(3), pp. 415-421.

Indraratna, B., Ngo, N. T. & Rujikiatkamjorn, C., 2011. Behaviour of geogrid-reinforced ballast under various levels of fouling. *Geotextiles and geomembranes,* Volume 29, pp. 313-322.

Infante, D., Martinez, G., Arrua, P. & Eberhardt, M., 2016. Shear strength behaviour of different geosynthetic reinforced soil structure from direct shear test. *International journal of geosynthetics and ground engineering,* 2(17).

Innovation in textiles, 2012. *Nonwovens in Asia: Big dimensions, big potential.* [Online] Available at: http://www.innovationintextiles.com/nonwovens-in-asia-big-dimensions-big-potential/ [Accessed 5 July 2016].

Jafari, N., Stark, T. & Merry, S., 2012. The July 10 2000 Payatas Landfill Slope Failure. *International journal of geoengineering case histories,* 2(3), pp. 208-230.

Kamon, M., Ali, Faisal Hj, Katsumi, Takeshi, Akai, Tomoyuki, Inui, Toru & Saranvanan, Mariappan, 2003. *Interface shear strength of composite landfill liner,* Kyoto, Japan: (京都大学) 0048 新制・課程 博士 博士(地球環境学) 甲第13255号 地環博第21号 新制/地環/4 UT51-2007-H528 2007-03-23 京 都大学大学院地球環境学舍地球環境学専攻 (主査)教授 嘉門 雅史, 教授 松井 三郎, 助教授 勝見 武 学位規則第4条第1項該当, Kyoto University.

Kamon, M., Mariappan, S., Katsumi, T., Inui, T. & Akai, T., 2008. *Large scale shear tests on interface shear performance of landfill liner systems.* Shanghai, Asian Regional Conference on geosynthetics.

Kavazanjian Jr., E. & Merry, S. M., 2005. *The 10 July 2000 Payatas landfill failure.* Cagliari, CISA, Environmental Sanitary Engineering Centre, Italy.

Kaytech Engineered Fabrics, 2010a. *Waste containment landfill lining Mooinooi, North West Province,* Mooinooi: Kaytech Engineered Fabrics.

Kaytech Engineered Fabrics, 2010b. *EnviroFix: Needlepunched and thermally locked geosynthetic clay liner*, Johannesburg: Kaytech engineered fabrics.

Kaytech Engineered Fabrics, 2012. *Miragrid GX Geogrids.* [Online]
Available at: http://kaytech.co.za/wp-content/uploads/2015/11/Miragrid-GX_k-Tech-Data.pdf
[Accessed 20 January 2017].

Kaytech Engineered Fabrics, 2014a. *GCLs: manufacturing quality and management manual*, Atlantis: Kaytech Engineered Fabrics.

Kaytech Engineered Fabrics, 2014b. *Envirotex provisional technical data sheet*, Johannesburg: Kaytech Engineered Fabrics.

Kaytech Engineered Fabrics, 2014c. *Bidim technical data sheet*, Johannesburg: Kaytech Engineered Fabrics.

Kaytech Engineered Fabrics, 2015. *Bidim: General civil engineering applications.* [Online]
Available at: http://kaytech.co.za/wp-content/uploads/2015/11/bidim-brochure-BR-GNRL-0637-04_2015.pdf [Accessed 3 March 2017].

Kaytech Engineered Fabrics, 2016. *Product details.* [Online]
Available at: http://kaytech.co.za/product/envirofix/ [Accessed 25 March 2016].

Kim, D., 2006. *Multi-scale assessment of geotextile-geomembrane interaction*, Georgia: PhD book. Georgia Institute of Technology.

Koelsch, F., 2000. *Stability problems of landfills - The Payatas landslide*, Braunschweig: Dr. Koelsch Geoenvironmental Technology LLC.

Koerner, R. M., 2005. Designing with geosynthetics: Fifth Edition. In: *Designing with geomembranes.* New Jersey: Pearson Education, Inc, pp. 428-630.

Koerner, R. M. & Hwu, B. L., 1991. Stability and tension considerations regarding cover soils on geomembrane lined slopes. *Geotextiles and geomemrbanes,* 10(4), pp. 335-355.

Koerner, R. M. & Soong, T. Y., 1998. *Analysis and design of veneer cover soils.* Georgia, Proceedings of the 6th IGS conference.

Landva, A. O. & Dickinson, S. J. E., 2012. Landslides in landfills. *ISSMGE Bulletin,* 6(1), pp. 10-18.

Lavigne, Franck; Heng, Mathias; Iskandarsyah, Yan, Wassmer, Patrick; Gomez, Christopher, Davies, Thimoty; Hadmoko, Danang; Gallard, JC & Fort, Monique; Texier, Pauline; Pratomo, Indyo, 2014. *The 21 February 2005, catastrophic waste avalanche at Leuwigajah dumpsite, Bandung, Indonesia,* Geoenvironmental Disasters: Springer.

Legg, P., 2010. *Final design and operating plan report for the extension of the Tutuka power station landfill site,* Benoni: Peter Legg Consulting: Geo-Environmental Engineers.

Ling, H. I. & Leshchinsky, D., 1997. Seismic stability and permanent displacement of landfill cover systems. *Geotechnical and geoenvironmental engineering, ASCE,* 123(2), pp. 113-122.

Liu, C.-N., Zornberg, J. G., Chen, Tsong-Chia; Ho, Yu-Hsien & Lin, Bo-Hung, 2009. Behaviour of geogrid-sand interface in direct shear mode. *Journal of geotechnical and geoenvironmental engineering,* pp. 1863-1871.

Mackey, R. & von Maubeuge, K., 2004. *Advances in Geosynthetic Clay Liner technology: 2nd Symposium, Issue 1456.* 2nd Symposium ed. West Conshohocken: ASTM International.

McCartney, J. S. & Zornberg, J. G., 2003. *Characterization of GCL shear strength variability.* Canada, 56th Canadian Geotechnical Conference.

McCartney, J. S., Zornberg, J. G. & Swan, R. H., 2002. *Internal and interface shear strength of Geosynthetic Clay Liners (GCLs),* Colorado: Department of Civil, Environmental and Architectural Engineering.

McCartney, J. S., Zornberg, J. G. & Swan, R. H., 2004. Effect of specimen conditioning on geosynthetic clay liner shear strength. *University of Texas at Austin,* pp. 635-643.

MIRAFI, 1996. *Geotextile filter design, application and product selection guide,* Pendergrass: Ten Cate nicolon.

Mitchell, J., Seed, R. & Seed, B., 1990. Kettleman Hills waste landfill slope failure: Liner-system properties. *Journal of geotechincal engineering,* Volume 116, pp. 647-668.

Munawar, E. & Fellner, J., 2013. *Guidelines for design and operation of municipal solid waste landfills in tropical climates,* Banda Aceh: International Solid Waste Association (ISWA).

Mwai, M., Wichuk, K. & McCartney, D., 2010. *Use of tire-derived aggregate (TDA) for leachate collection and drainage systems.* [Online]
Available at: https://www.slideshare.net/SWANANLC/implications-of-using-tirederived-aggregate-for-landfill-leachate-collection-drainage-systems [Accessed 15 July 2017].

Orebowale, P. B., 2006. *Investigating the stability of geosynthetic landfill capping systems,* Loughborough: Loughborough University, PhD book.

Padade, A. H. & Mandal, J. N., 2012. *Direct shear test on expanded polystyrene (EPS) geofoam.* Valencia, 5th European Geosynthetic Congress.

Parra, D., Valdivia, R. & Soto, C., 2012. *Analysis of shear strength non-linear envelopes of soil-geomembrane interface and its influence in the heal leach pad stability.* Lima, Peru, Second Pan American Geosynthetics Conference & Exhibition GeoAmericas 2012.

Pollux waste to energy, 2016. *Types of landfill failures.* [Online]
Available at: http://www.polluxconsulting.com/stability/slope-stability/types-of-landfill-failures [Accessed 16 June 2016].

Qian, X., 2008a. Critical interfaces in geosynthetic multilayer liner system of a landfill. *Water science and engineering,* 1(4), pp. 22-35.

Qian, X., 2008b. Limit equilibrium analysis of translational failure of landfills under different leachate buildup conditions. *Water Science and Engineering,* 1(1), pp. 44-62.

Qian, X. & Koerner, R., 2014. Critical interfaces and waste placement in landfill design. *Institution of civil engineers.*

Qian, X. & Koerner, R. M., 2004. Effect of apparent cohesion on translational failure analyses of landfills. *Journal of geotechnical and geoenvironmental engineering,* 130(1), pp. 71-80.

Qian, X., Koerner, R. M. & Gray, D. H., 2003. Translational failure analysis of landfills. *Journal of geotechnical and geoenvironmental engineering,* 129(6), pp. 506-519.

Reddy, K. & Basha, M., 2014. *Slope stability of waste dumps and landfills: State-of-the-art and future challenges.* Kakinada, Indian Geotechnical Conference IGC.

Reddy, K. R., Kosgi, S. & Motan, S., 1996. Interface shear behaviour of landfill composite liner systems: A finite element analysis. *Geosynthetics International,* 3(2), pp. 247-275.

Rouncivell, W., 2007. *Experimental investigation of the shear strength characteristics of a geosynthetic clay liner and its application in a local landfill lining system,* Cape Town: University of Cape Town Masters book.

Russell, D., Jones, V. & Dixon, N., 1998. Shear strength properties of geomembrane/geotextile interfaces. *Geotextiles and geomembranes,* Volume 16, pp. 45-71.

Seo, M. W., Park, J. B., Park, I. J. & Cho, N. J., 2003. Development of strain-softening modeling for interfaces between geosynthetics. *한국토목섬유학회논문집,* 2(1), pp. 57-68.

Shukla, S. K. & Yin, J.-H., 2006. *Fundamentals of geosynthetic engineering.* The Netherlands: Taylor & Francis.

Simpson, M. & Siebken, J., 1997. *A comparison of High Density Polyethylene (HDPE) and Linear Low Density Polyethylene (LLDPE) geomembranes.* Bangalore, India, Geosynthetics Asia.

Stark, T. D., Williamson, T. A. & Eid, H. T., 1996. HDPE geomembrane/geotextile interface shear strength. *Journal of geotechnical engineering,* pp. 197 - 203.

Tancott, G., 2013a. Ensuring environmental safety with landfill liner. *Infrastructure news and service delivery,* 14 Novemeber.

Tancott, G., 2013b. *Kaytech's Bidim solves groundwater drainage problem.* [Online]
Available at: https://www.infrastructurene.ws/2013/10/16/kaytechs-bidim-solves-groundwater-
drainage-problem/# [Accessed 5 March 2017].

Tencate, 2013. *Miragrid - Stable and secure soil reinforcement,* Johannesburg: Kaytech.

Thiel, R., 2001. *Peak vs residual shear strength for landfill bottom liner stability analyses,* CA, USA:
Oregon House.

Thiel, R. S. & Criley, K., 2005. *Hydraulic conductivity of partially prehydrated GCLs under high
effective confining stresses for three real leachates.* Texas, Proceedings of the Geo-Frontiers 2005
Conference in Austin.

TRI Australasia, 2013. *Interface friction.* [Online] Available at: http://tri-env.com.au/fiction/
[Accessed 12 June 2017].

Triplett, E. J. & Fox, P. J., 2001. Shear strength of HDPE geomembrane/geosynthetic clay liner
interfaces. *Journal of geotechnical and geoenvironmental engineering,* Issue 127, pp. 543-552.

Tuna, S. C. & Altun, S., 2012. Mechanical behaviour of sand-geotextile interface. *Scientia Iranica,*
19(4), pp. 1044-1051.

United States Plastic Corparation, 2008. *What are the differences between HDPE, LDPE, XLPE,
LLDPE and UHMWPE?.* [Online]
Available at: http://www.usplastic.com/knowledgebase/article.aspx?contentkey=508
[Accessed 12 April 2017].

Wu, H.-m., Shu, Y.-m. & Zhu, J.-g., 2011. Implementation and verification of interface constitutive
model in FLAC 3D. *Water science and engineering,* 4(3), pp. 305-316.

Zaini, M. I., Kasa, A. & Nayan, A. M., 2012. Interface shear strength of geosynthetic clay liner (GCL)
and residual soil. *International journal on advanced science engineering information technology,* 2(2),
pp. 43-45.

Zanzinger, H., 2012. *Good-better-best HDPE Geomembranes in Engineered Geoenvironmental Applications,* Wurzburg, Germany: SKZ - German Plastics Center.

Zornberg, J. G., McCartney, J. S. & Swan Jr., R. H., 2005. Analysis of a Large Database of GCL Internal Shear Strength Results. *Journal of geotechnical and geoenvironmental engineering,* 131(3), pp. 368-380.

www.ingramcontent.com/pod-product-compliance
Lightning Source LLC
LaVergne TN
LVHW042110190726
843493LV00006B/1433